THE PROCEEDINGS OF THE FESTSCHRIFT IN HONOR OF

BRUCE H J MCKELLAR
AND
GIRISH C JOSHI

THE PROCEEDINGS OF THE FESTSCHRIFT IN HONOR OF

BRUCE H J MCKELLAR
AND
GIRISH C JOSHI

The University of Melbourne, Australia 29 – 30 November 2006

Editor

Raymond R Volkas

The University of Melbourne, Australia

World Scientific

NEW JERSEY • LONDON • SINGAPORE • BEIJING • SHANGHAI • HONG KONG • TAIPEI • CHENNAI

Published by

World Scientific Publishing Co. Pte. Ltd.

5 Toh Tuck Link, Singapore 596224

USA office: 27 Warren Street, Suite 401-402, Hackensack, NJ 07601

UK office: 57 Shelton Street, Covent Garden, London WC2H 9HE

British Library Cataloguing-in-Publication Data
A catalogue record for this book is available from the British Library.

THE PROCEEDINGS OF THE FESTSCHRIFT IN HONOR OF BRUCE H J MCKELLAR AND GIRISH C JOSHI

ISBN-13 978-981-270-854-0 (pbk)
ISBN-10 981-270-854-5 (pbk)

Printed in Singapore.

PREFACE

The *Festschrift in Honour of Bruce McKellar and Girish Joshi* was held in the School of Physics at the University of Melbourne in Victoria, Australia on 29-30 November 2006. Bruce and Girish have had long and distinguished careers as researchers and academics, and it was my very great pleasure to host this celebration of their careers and achievements. I also wish to personally thank them for their mentoring over the years as I progressed through my own career. Special thanks from me are due to Girish, who was my thesis advisor and always full of ideas about the cutting edge of theoretical particle physics research.

Bruce McKellar received his PhD from the University of Sydney in 1966 under S. T. Butler, R. M. May and M. A. Naqvi. He was then a Fulbright Fellow at the Institute for Advanced Study in Princeton, before moving back to Sydney as a Lecturer and then Senior Lecturer. He was appointed Professor of Theoretical Physics at Melbourne in 1972 at an enviably young age, and maintained that position until his retirement in 2007. He also served for a time as Head of the School of Physics and spent several years as Dean of the Faculty of Science. Most recently, he has been active in senior leadership roles within the Australian Academy of Science, first as Secretary for Physical Sciences and most recently as Foreign Secretary. Amongst many honours and awards, Bruce received the 2006 Massey Medal from the Institute of Physics (U.K.) for his contributions to the standard model of particle physics.

His contributions to research cover a wide range of areas, with an emphasis on theoretical particle and nuclear physics. Work on parity violation in nuclear physics and on three-body forces was the first application of current algebra and chiral symmetry techniques in nuclear physics, and was a precursor to the currently fashionable applications of chiral perturbation theory to the interactions between nucleons. With B. Gibson, he did the standard calculation of non-mesonic decays of Λ-hypernuclei in the standard model. With S. R. Choudhury, X.-G. He and S. Pakvasa, he did the definitive calculation of the neutron electric dipole moment, and this led to extensive work on CP violation beyond the standard model. His work with M. Thomson on neutrino kinetic equations is now regularly used in early universe and supernovae calculations.

Girish Joshi graduated with a PhD from Delhi University in 1965 under the supervision of R. C. Majumdar. He was then a C.S.I.R. postdoctoral fellow at Delhi, before moving to the University of Manchester as a Lecturer. He was also a Visiting Scientist at the International Centre for Theoretical Physics in Trieste, Italy, and prior to taking up his Melbourne appointment in 1970 he was a postdoctoral fellow at Rutgers University in the United States. He retired at the end of 2005 as a Reader in Physics. He has made important contributions to Regge theory, analytic continuation in the complex angular momentum plane, pentaquark hadrons and dibaryons, physics beyond the standard model and B-physics phenomenology. His particular passion has been the exploration of novel mathematical structures such as quaternions, octonions, Jordan algebras and the like in physics.

Both Bruce and Girish have been outstanding advisors to dozens of successful doctoral and Honours students.

The meeting in November consisted of plenary talks about particle physics and cognate areas, given by some of Bruce's and Girish's closest associates, collaborators, friends and former students. I thank all who participated, speakers and listeners, for making this meeting a success.

It is with great sadness that I have to note the premature passing of Professor Rev. Ron Anderson of Boston College, Girish's first PhD student and a lively participant at the Festschrift.

I thank Professor Herb Fried of Brown University for contributing to the written Festschrift despite being unable to travel to Australia for the meeting. Thanks also to Professor R. Rajaraman of Jawaharlal Nehru University who presented a fascinating public lecture on nuclear disarmament in South Asia in conjunction with the Festschrift. Professor S. R. Choudhury of Delhi University, a frequent collaborator of both Bruce and Girish, was unfortunately unable to attend due to circumstances beyond his control, but was very much there in spirit.

The meeting would not have been a success without the organisational skills of Helen Conley, Executive Manager of the School of Physics at Melbourne, and her team of Cilla Gloger (general organisation and catering), Marcia Damjanovic-Napoleon (registration and finance), Tim Dyce (IT support) and Janet Carlon (group photograph). Thanks to Sandy Law for maintaining the Festschrift web site, and his dogged pursuit of the candid photograph.

Special thanks to Professor David Caro for officially opening the meeting. Professor Caro, a past Vice-Chancellor of the University of Melbourne, was Head of the School of Physics when Bruce and Girish were appointed. His vision enabled the creation of a substantial theoretical particle physics group at Melbourne.

Finally, many of us know that physics is both vocation and avocation. "Retirement" is but a new beginning. You shall continue to see McKellar and Joshi on the pages of your favourite journal.

Raymond R. Volkas
(Chair, Festschrift in Honour of Bruce McKellar and Girish Joshi)

Melbourne, Australia
July 2007

PHOTOGRAPHS

Bruce McKellar

Girish Joshi

Geoff Taylor

Jerry Stephenson

Ron Anderson

Roy Volkas, Bruce McKellar, Girish Joshi, David Caro and Geoff Taylor

Bruce Barrett, Bruce McKellar and Sid Coon

Robert Foot, Ray Volkas, Girish Joshi and Ron Anderson

Girish Joshi, David Caro and Xiao-Gang He

Families of McKellar and Joshi

Bruce McKellar and Doug Rajaraman

Jeanette Fyffe, James McCaw, Kristian McDonald and friend

CONTENTS

---

*Speaker.

THREE NUCLEON FORCES AND NUCLEAR STRUCTURE

B. R. BARRETT*

Physics Department, University of Arizona,
Tucson, Arizona 85721-0081, USA
**E-mail: bbarrett@physics.arizona.edu*
www.physics.arizona.edu

The author's earlier research with Professor Bruce McKellar, some 30 years ago, led to the Tucson-Melbourne three-nucleon interaction, which is still relevant today. The new significance of three-nucleon forces in determining the structure of atomic nuclei will be discussed. This has led to increased efforts to learn more about the nature of these three-nucleon interactions, both experimentally and theoretically. The recently developed no-core shell model (NCSM) has the ability to test different theoretical models for three-nucleon forces by making direct comparisons of results produced by these forces with experimental data.

Keywords: nuclear structure, Tucson-Melbourne three-nucleon interaction, no-core shell model.

1. Introduction

Congratulations to Bruce McKellar and Girish Joshi on their retirements from the University of Melbourne! It is indeed a great pleasure for me to take part in this retirement Festschrift in their honor, because I have known Bruce McKellar for 40 years and have had a connection with the University of Melbourne since 1973.

My most memorable and significant connection is related to the topic of three-nucleon (NNN) interactions. In the early 1970s I taught a graduate level course on nuclear many-body theory at the University of Arizona, which was being audited by Dr. Sidney Coon, who was a research associate in our group at that time. When I got to the part regarding the theory of the contribution of NNN interactions to the binding energy of nuclear matter, Dr. Coon suggested that we take up this problem with Dr. Michael Scadron, also in our department. The basic idea was to include the pion-nucleon scattering amplitude into the standard two-pion exchange NNN interaction of the Fujita-Miyazawa[1] structure. The results of this investigation[2] were published in 1975.

Dr. Scadron was also a collaborator of Bruce McKellar and, hence, took up this problem with him, which produced a number of improvements and refinements, leading to what is known today as the Tucson-Melbourne three-nucleon (TM NNN) interaction, which is one of the most-recognized NNN interactions worldwide. Our paper[3] on the TM NNN interaction in 1979 has become a widely cited publication, still being referenced today, because of the now-recognized importance of NNN interactions to the binding energies and other properties of finite atomic nuclei as well as of nuclear matter. The latest version of the TM NNN interaction has been developed by Coon and collaborators,[4] with input from Chiral Effective Field Theory[5] (χEFT).

2. The Significance of Three-Nucleon Interactions

The importance of NNN interactions to the theory of nuclear structure has become more apparent within the last ten years with tremendously improved procedures for solving the nuclear many-body problem, such as, the Green Function Monte Carlo[6,7] (GFMC) and No-Core Shell Model[8,9] (NCSM) approaches. In a ground-breaking benchmark investigation[10] in 2001, seven nuclear-theory groups (totally 17 theorists!) calculated the properties of the $A = 4$ system using the same nucleon-nucleon (NN) potential, *i.e.*, the Argonne V8′ (AV8′) potential.[6] Within the limits of error of the seven different approaches, all the groups obtained the same result for the binding energy of ^{4}He. The most significant result of this investigation is that nuclear-structure theory has now progressed to the stage, where truly accurate and meaningful calculations can be performed for the properties of light nuclei, given your choice for the NN interaction. The bad news is that the mean result obtained for the binding energy underbinds ^{4}He by about 2.3 MeV. However, the positive conclusion to be drawn from this negative result is that the discrepancy between theory and experiment has real physical significance, because the theoretical results are accurate and meaningful. Studies[6–9,11] for other light nuclei from $A = 5$ to $A = 12$ show a consistent underbinding of nuclei, compared with experiment, when only local NN interactions are utilized. Consequently, something is actually missing in the theory of nuclear structure, when only local NN interactions are employed. The logical choice for the missing ingredient is NNN interactions, since it is well-known that such forces naturally arise theoretically, when the interaction among nucleons is derived from χEFT.[5,12–14] This has led to a great interest, both theoretically and experimentally, in obtaining as much information as possible about the nature of these NNN interactions.

3. The No-Core Shell Model (NCSM)

In the last 15 years a new *ab initio* microscopic approach has been formulated for calculating the properties of atomic nuclei, known as the No-Core Shell Model[8,9] (NCSM). In the NCSM approach, there is no inert, closed core of nucleons, surrounded by a few valence nucleons;[15,16] but all A nucleons are taken as being active. This new procedure has a number of advantages over the older, standard shell model,[16] such as avoiding the intruder-state effect; yielding an exact, converged result for the effective interaction in a given model space; providing energies relative to the vacuum; and allowing for an exact treatment of the spurious center-of-mass (CM) motion problem.

The NCSM formalism has been given in earlier publications[8,9] and will not be repeated here. The reader is referred to the literature for a detailed description of this approach, which can be summarized as follows:

(1) Start with the A-nucleon Hamiltonian in relative coordinates and your choice for the NN interaction (or NN + NNN interactions).

(2) Bind the CM of the nucleus in an harmonic-oscillator (HO) potential well,[17] *i.e.,* add the CM HO potential to the Hamiltonian in part (1). This does not change any of the intrinsic energies but provides a single-particle HO basis for the performing of numerical calculations. It also confines the nucleons in a HO potential well, which aids in the convergence of the final numerical results.

(3) In order to solve the nuclear Schrödinger equation for the A-nucleon Hamiltonian, truncate to a finite model space. This produces a tractable numerical calculation but converts the initial NN (or NN + NNN) interactions into effective A-nucleon interactions.

(4) Because one cannot, in general, solve the A-nucleon problem with A-nucleon interactions (*i.e.,* for $A > 4$), it is necessary to restrict the calculations to effective NN (or effective NNN) interactions and operators, known as the two-body cluster (or three-body cluster) approximation.

(5) Pick a given cluster level, *e.g.,* the two-body cluster, and solve the Schrödinger equation at this level, *i.e.,* for the eigenenergies and eigenfunctions of two nucleons in an HO potential well interacting through an NN interaction.

(6) Pick the size of the model space, in which you want to perform your calculations (*i.e.,* in terms of the eigenstates up to a given energy above the ground-state energy), and then use the solutions in part (5) to calculate exactly the two-body effective interaction in this model space, as well as other two-body effective operators, using the unitary transformation method.[18–22]

(7) Finally, by diagonalization, solve the Schrödinger equation in the truncated model space for all A nucleons active, using the effective interaction in part (6). Repeat the calculation for increasing size of the model space to check the convergence of the numerical results. There is also a pseudo-dependence on the HO energy, $\hbar\Omega$, due to the truncation of the Hilbert space, but this dependence goes away as the size of the model space approaches infinity.

4. Testing Different Theoretical NNN Interaction Models

How all of this becomes relevant to the earlier discussion of NNN interactions is the fact that the above procedure can be repeated at the three-body cluster level, including NNN interactions. Our present knowledge regarding the physical nature of NNN interactions is based purely on theoretical models, such as the TM,[4] the Urbana-IX (UIX) and the Illinois-2 (IL2).[7,23] One of the main problems is that most numerical calculations involve NNN interactions, which are *not* theoretically self-consistent with the accompanying NN interactions, such as AV18 + UIX or AV8′

+ TM. In these cases the unknown paramenters in the NNN-interaction models are adjusted to fit some data for light nuclei, such as, the binding energies of ^{3}H and ^{4}He, calculated with the NN + NNN interactions. These NN + fitted-NNN interactions are then used to calculate the properties of heavier nuclei.

In practice, one wants to utilize interactions among the nucleons based on the QCD Lagrangian. This is now possible in terms of χEFT, in which NNN interactions consistent with the accompanying NN interactions are generated as one calculates higher-order terms in the χEFT expansion.[5,13] Figure 1 shows the first three terms, which occur for the NNN interaction in this expansion at next-to-next-to-lowest order (N^{2}LO). The first term is related to the Fujita-Miyazawa two-pion exchange

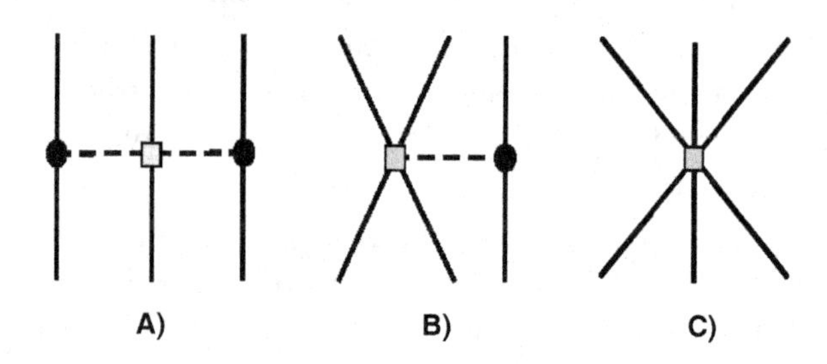

Fig. 1. The first three terms in the NNN interaction, occurring in N^{2}LO of the χEFT expansion.

contribution[1,5] to the NNN interaction. The second and third processes contain what are known as contact terms, which involve low-energy constants (LECs). At some future time these LECs will need to be determined from Lattice Gauge calculations, but for now they must be obtained by fitting to experimental data. In the case of the Idaho χEFT interaction,[24] there are two choices for these LECs, which describe the experimental data equally well.[25] Consequently, a test is needed for deciding between these two choices for the LECs, or, in general, among any set of different theoretcial models for NNN interactions.

Such a test exists in the form of the NCSM, which allows nuclear-structure calculations at the three-body cluster level, including NNN forces. Because of the sensitive dependence of theoretical NNN interactions on spin and isospin, such forces can have a dramatic effect not only on the binding energies of nuclei but also on their spectra. For example, the calculated ground-state (I$^\pi$,T) for ^{10}B is $(1^+, 0)$ using only local NN interactions,[26] while it is the correct value of $(3^+, 0)$, when NNN interactions are included in the calculations.[7,9]

Thus, by performing NCSM calculations at the three-body cluster level with different theoretical models for the NNN interactions and comparing the results with experiment, one has a test for these theoretical models. Preliminary calculations[25,27] of this kind have been performed for ^{6}Li and ^{7}Li and are shown in Figs. 2 and 3, respectively. The 3NF-A and 3NF-B results in these two figures refer to the two equally good choices for the LECs in Fig. 1, *i.e.*, terms B) and C). At this time, the purpose is *not* to make an argument as to whether choice 3NF-B is better than

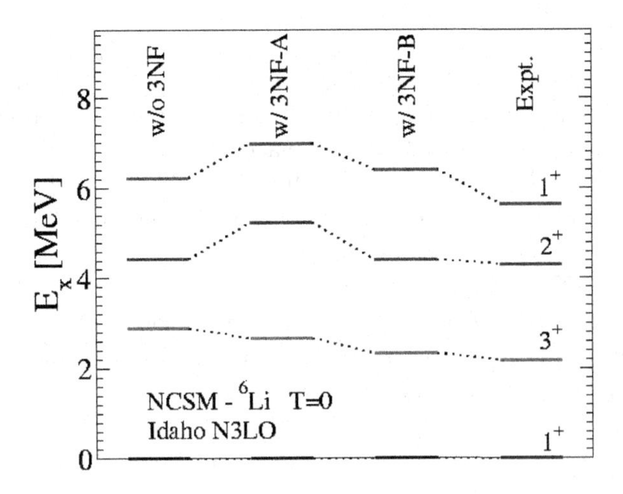

Fig. 2. Dependence of the excitation energies of the lowest states of ^{6}Li on the interaction.[27] Results with the NN interaction only and with the 3NF-A and 3NF-B included (see text for details) are compared with the experimental values.

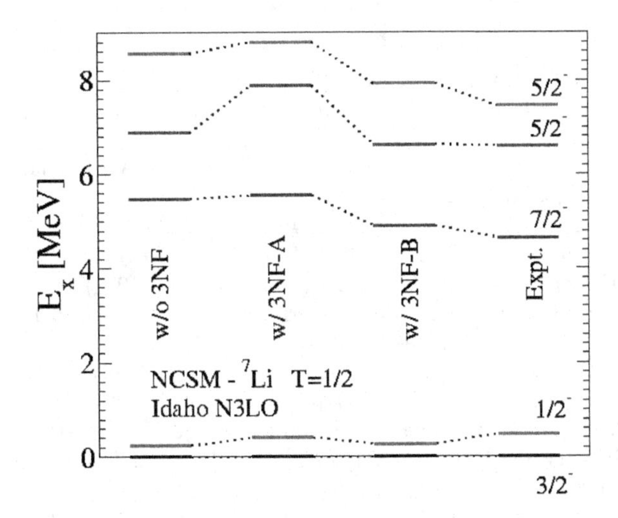

Fig. 3. Dependence of the excitation energies of the lowest states of ^{7}Li on the interaction.[25] Results with the NN interaction only and with the 3NF-A and 3NF-B included (see text for details) are compared with the experimental values.

choice 3NF-A, based on these limited calculations for ^{6}Li and ^{7}Li, but to illustrate how the NCSM can be employed to make such evaluations. Clearly, more extensive calculations for many more nuclei are necessary before one can meaningfully evaluate and rate different NNN interactions.

5. Conclusions

Because NNN interactions are important for understanding the details of nuclear structure and because all current models of such forces are purely theoretical, there is a real need for a procedure for evaluating the accuracy of these different theoretical models. The NCSM provides such a procedure, because of its ability to perform nuclear-structure calculations at the three-body cluster level, including NNN interactions.

In conclusion, NNN interactions are as interesting and timely an area of research in nuclear-structure theory today, as they were some 30 years ago, when Bruce McKellar and I first collaborated, leading to a long, enjoyable and productive scientific relationship as well as friendship.

Acknowledgments

B.R.B. wishes to thank A. F. Lisetskiy for assistance with the figures and the Service de Physique Théorique, CEA, Saclay, France, and TRIUMF, Vancouver, BC, Canada, for their hospitality during the preparation of this manuscript. This research was partially supported by NSF grant PHY-0555396.

References

1. J. Fujita and H. Miyazawa, *Prog. Theor. Phys.* **17**, 360 (1957).
2. S. A. Coon, M. D. Scadron and B. R. Barrett, *Nucl. Phys.* **A242**, 467 (1975); **A254**, 542 (1975).
3. S. A. Coon, M. D. Scadron, P. C. McNamee, B. R. Barrett, D. W. E. Blatt and B. H. J. McKellar, *Nucl. Phys.* **A317**, 242 (1979).
4. S. A. Coon and M. T. Pena, *Phys. Rev. C* **48**, 2559 (1993), nucl-th/9305017; S. A. Coon and H. K. Han, *Few Body Syst.* **30**, 131 (2001), nucl-th/0101003.
5. J. L. Friar, D. Hüber and U. van Kolck, *Phys. Rev.* **59**, 53 (1999).
6. B. S. Pudliner, V. R. Pandharipande, J. Carlson, S. C. Pieper and R. B. Wiringa, *Phys. Rev. C* **56**, 1720 (1997), nucl-th/9705009.
7. S. C. Pieper and R. B. Wiringa, *Ann. Rev. Nucl. Part. Sci.* **51**, 53 (2001), nucl-th/0103005; S. C. Pieper, K. Varga and R. B. Wiringa, *Phys. Rev. C* **66**, 044310 (2002); S. C. Pieper, R. B. Wiringa and J. Carlson, *Phys. Rev. C* **70**, 054325 (2004).
8. P. Navrátil, J. P. Vary and B. R. Barrett, *Phys. Rev. Lett.* **84**, 5728 (2000); *Phys. Rev. C* **62**, 054311 (2000).
9. P. Navrátil and W. E. Ormand, *Phys. Rev. Lett.* **88**, 152502 (2002); *Phys. Rev. C* **68**, 034305 (2003), nucl-th/0305090.
10. H. Kamada, *et al.*, *Phys. Rev. C* **64**, 044001 (2001), nucl-th/0104057.
11. P. Navrátil and E. Caurier, *Phys. Rev. C* **69**, 014311 (2004).
12. S. Weinberg, *Phys. Lett. B* **251**, 288 (1990); *Nucl. Phys. B* **363**, 3 (1991).
13. U. van Kolck, *Phys. Rev. C* **49**, 2932 (1994).
14. C. Ordonez, L. Ray and U. van Kolck, *Phys. Rev. Lett.* **72**, 1982 (1994); *Phys. Rev. C* **53**, 2086 (1996), hep-ph/9511380.
15. E. Caurier, G. Martinez-Pinedo, F. Nowacki, A. Poves and A. P. Zuker, *Rev. Mod. Phys.* **77**, 427 (2005).
16. B. R. Barrett, *Czech. J. Phys.* **49**, 1 (1999).

17. H. Lipkin, *Phys. Rev.* **109**, 2071 (1958).
18. S. Okubo, *Prog. Theor. Phys.* **12**, 603 (1954).
19. J. Da Providencia and C. M. Shakin, *Ann. Phys. (N.Y.)* **30**, 95 (1964).
20. K. Suzuki and S. Y. Lee, *Prog. Theor. Phys.* **64**, 2091 (1980).
21. K. Suzuki, *Prog. Theor. Phys.* **68**, 246 (1982); *op cit.*, 1999 (1982).
22. K. Suzuki and R. Okamoto, *Prog. Theor. Phys.* **70**, 439 (1983); *ibid.*, **92**, 1045 (1994).
23. S. C. Pieper, V. R. Pandaripande, R. B. Wiringa and J. Carlson, *Phys. Rev. C* 64, 014001 (2001), nucl-th/0102004.
24. D. R. Entem and R. Machleidt, *Phys. Rev. C* **68**, 041001(R) (2003), nuth-th/0304018.
25. A. Nogga, P. Navrátil, B. R. Barrett and J. P. Vary, *Phys. Rev. C* **73**, 064002 (2006).
26. E. Caurier, P. Navrátil, W. E. Ormand and J. P. Vary, *Phys. Rev. C* **66**, 024314 (2002).
27. A. Nogga, E. Epelbaum, P. Navrátil, W. Glöckle, H. Kamada, Ulf-G. Meissner, H. Witala, B. R. Barrett and J. P. Vary, *Nucl. Phys.* **A737**, 236 (2004).

HOW MAGNETIC IS THE NEUTRINO? *

N. F. BELL†

School of Physics,
The University of Melbourne,
Victoria, 3010, Australia
†*E-mail: n.bell@physics.unimelb.edu.au*

The existence of a neutrino magnetic moment implies contributions to the neutrino mass via radiative corrections. We derive model-independent "naturalness" upper bounds on the magnetic moments of Dirac and Majorana neutrinos, generated by physics above the electroweak scale. For Dirac neutrinos, the bound is several orders of magnitude more stringent than present experimental limits. However, for Majorana neutrinos the magnetic moment bounds are weaker than present experimental limits if μ_ν is generated by new physics at $\sim$ 1 TeV, and surpass current experimental sensitivity only for new physics scales > 10 – 100 TeV. The discovery of a neutrino magnetic moment near present limits would thus signify that neutrinos are Majorana particles.

Keywords: Neutrino, magnetic moment, neutrino mass.

1. Introduction

In the Standard Model (minimally extended to include non-zero neutrino mass) the neutrino magnetic moment is non-zero, but small, and is given by[1]

$$\mu_\nu \approx 3 \times 10^{-19} \left(\frac{m_\nu}{1\text{eV}}\right) \mu_B, \tag{1}$$

where m_ν is the neutrino mass and μ_B is the Bohr magneton. An experimental observation of a magnetic moment larger than that given in Eq.(1) would thus be a clear indication of physics beyond the minimally extended Standard Model. Current laboratory limits are determined via neutrino-electron scattering at low energies, with $\mu_\nu < 1.5\times 10^{-10}\mu_B$[2] and $\mu_\nu < 0.7\times 10^{-10}\mu_B$[3] obtained from solar and reactor experiments, respectively. A stronger limit can be obtained from constraints on energy loss from stars, $\mu_\nu < 3 \times 10^{-12}\mu_B$.[4]

It is possible to write down a simple relationship between the size of the neutrino mass and neutrino magnetic moment. If a magnetic moment is generated by physics beyond the Standard Model (SM) at an energy scale Λ, as in Fig. 1a, we can generically express its value as

$$\mu_\nu \sim \frac{eG}{\Lambda}, \tag{2}$$

*This article is based upon the results of Refs. 10 and 12.

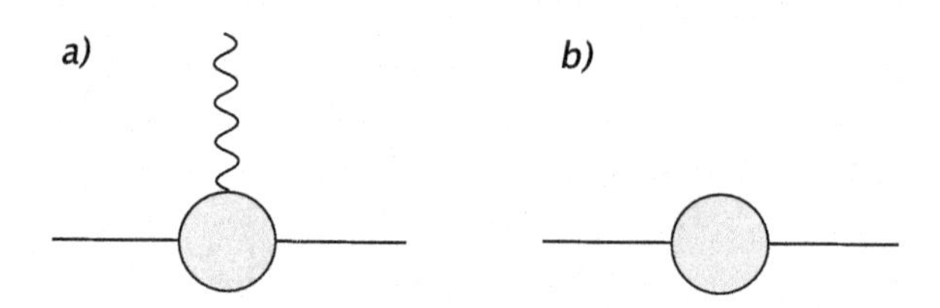

Fig. 1. a) Generic contribution to the neutrino magnetic moment induced by physics beyond the standard model. b) Corresponding contribution to the neutrino mass. The solid and wavy lines correspond to neutrinos and photons respectively, while the shaded circle denotes physics beyond the SM.

where e is the electric charge and G contains a combination of coupling constants and loop factors. Removing the photon from the same diagram (Fig. 1b) gives a contribution to the neutrino mass of order

$$m_\nu \sim G\Lambda. \tag{3}$$

We thus have the relationship

$$m_\nu \sim \frac{\Lambda^2}{2m_e}\frac{\mu_\nu}{\mu_B} \sim \frac{\mu_\nu}{10^{-18}\mu_B}[\Lambda(\text{TeV})]^2 \text{ eV}, \tag{4}$$

which implies that it is difficult to simultaneously reconcile a small neutrino mass and a large magnetic moment.

However, it is well known that the naïve restriction given in Eq.(4) can be overcome via a careful choice for the new physics. For example, we may impose a symmetry to enforce $m_\nu = 0$ while allowing a non-zero value for μ_ν,[5–8] or employ a spin suppression mechanism to keep m_ν small.[9] Note though, that these symmetries are typically broken by Standard Model interactions. By calculating contributions to m_ν generated by SM radiative corrections involving the magnetic moment interaction, we may thus obtain general, "naturalness" upper limits on the size of neutrino magnetic moments.

One possibility for allowing a large μ_ν while keeping m_ν small is due to Voloshin.[5] The original version of this mechanism involved imposing an $SU(2)_\nu$ symmetry, under which the left-handed neutrino and antineutrino (ν and ν^c) transform as a doublet. The Dirac mass term transforms as a triplet under this symmetry and is thus forbidden, while the magnetic moment term is allowed as it transforms as a singlet. However, the $SU(2)_\nu$ symmetry is violated by SM gauge interactions. For Majorana neutrinos, the Voloshin mechanism may be implemented using flavor symmetries, such as those in Refs. 6–8. These flavor symmetries are not broken by SM gauge interactions but are instead violated by SM Yukawa interactions.[a]

[a]We assume that the charged leptons masses are generated via the standard mechanism through Yukawa couplings to the SM Higgs boson. If the charged lepton masses are generated via a non-standard mechanism, SM Yukawa interactions do not necessarily violate flavor symmetries. However, such flavor symmetries must always be broken via some mechanism in order to obtain non-degenerate masses for the charged leptons.

Below, we shall estimate the contribution to m_ν generated by SM radiative corrections involving the magnetic moment term. This allows us to set general, "naturalness" upper limits on the size of neutrino magnetic moments. For Dirac neutrinos, these limits are several orders of magnitude stronger than present experimental bounds.[10] For Majorana neutrinos, however, the bounds are weaker.[11,12]

2. Dirac Neutrinos

We assume that the magnetic moment is generated by physics beyond the SM at an energy scale Λ above the electroweak scale. In order to be completely model independent, the new physics will be left unspecified and we shall work exclusively with dimension $D \geq 4$ operators involving only SM fields, obtained by integrating out the physics above the scale Λ. We thus consider an effective theory that is valid below the scale Λ, respects the $SU(2)_L \times U(1)_Y$ symmetry of the SM, and contains only SM fields charged under these gauge groups.

We start by constructing the most general operators that could give rise to a magnetic moment operator, $\bar{\nu}_L \sigma^{\mu\nu} F_{\mu\nu} \nu_R$. Demanding invariance under the SM gauge group, we have the following 6D operators

$$\mathcal{O}_B^{(6)} = \frac{g'}{\Lambda^2} \bar{L} \tilde{\phi} \sigma^{\mu\nu} \nu_R B_{\mu\nu} \,, \qquad \mathcal{O}_W^{(6)} = \frac{g}{\Lambda^2} \bar{L} \tau^a \tilde{\phi} \sigma^{\mu\nu} \nu_R W_{\mu\nu}^a \,. \tag{5}$$

where $B_{\mu\nu} = \partial_\mu B_\nu - \partial_\nu B_\mu$ and $W_{\mu\nu}^a = \partial_\mu W_\nu^a - \partial_\nu W_\mu^a - g\epsilon_{abc} W_\mu^b W_\nu^c$ are the $\mathrm{U}(1)_Y$ and $\mathrm{SU}(2)_L$ field strength tensors, respectively, and g' and g are the corresponding couplings. The Higgs and left-handed lepton doublet fields are denoted ϕ and L, respectively, and $\tilde{\phi} = i\tau_2 \phi^*$.

After spontaneous symmetry breaking, both $\mathcal{O}_B^{(6)}$ and $\mathcal{O}_W^{(6)}$ contribute to the magnetic moment. Through loop diagrams these operators will generate contributions to the neutrino mass. For example, the diagram in Fig. 2 will generate a contribution to the neutrino mass operator, $\mathcal{O}_M^{(4)} = \bar{L}\tilde{\phi}\nu_R$. Using dimensional analysis, we estimate[10]

$$m_\nu \sim \frac{\alpha}{16\pi} \frac{\Lambda^2}{m_e} \frac{\mu_\nu}{\mu_B} \,, \tag{6}$$

and thus

$$\mu_\nu \lesssim 3 \times 10^{-15} \mu_B \left(\frac{m_\nu}{1\ \mathrm{eV}}\right) \left(\frac{1\ \mathrm{TeV}}{\Lambda}\right)^2 \,. \tag{7}$$

If we take $\Lambda \simeq 1$ TeV and $m_\nu \lesssim 0.3$ eV, we obtain the limit $\mu_\nu \lesssim 10^{-15} \mu_B$, which is several orders of magnitude stronger than current experimental constraints. Given the quadratic dependence upon Λ, this constraint becomes extremely stringent for Λ significantly above the electroweak scale.

However, if Λ is not significantly larger that the EW scale, higher dimension operators are important, and their contribution to m_ν can be calculated in a model

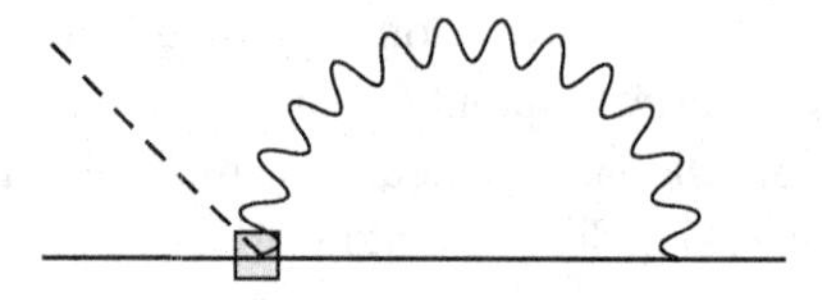

Fig. 2. Contribution to the 4D mass operator $\mathcal{O}_M^{(4)}$ due to insertions of the magnetic moment operators $\mathcal{O}_{B,W}^{(5)}$.

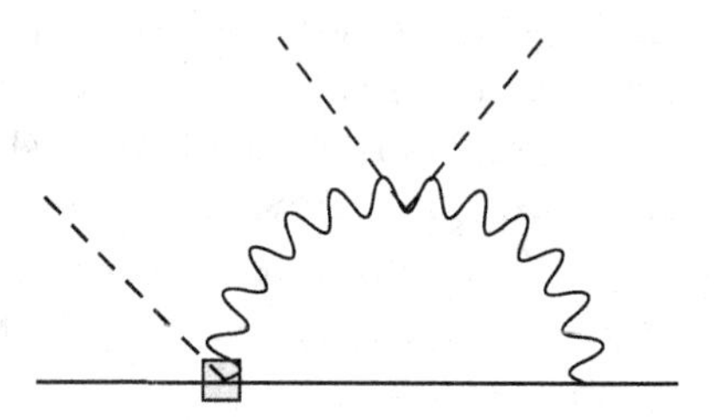

Fig. 3. Renormalization of the mass operator, $\mathcal{O}_M^{(6)}$, due to insertions of $\mathcal{O}_{B,W}^{(6)}$.

independent way. Through renormalization, both $\mathcal{O}_B^{(6)}$ and $\mathcal{O}_W^{(6)}$ will generate a contribution to the 6D neutrino mass operator

$$\mathcal{O}_M^{(6)} = \frac{1}{\Lambda^2}\bar{L}\tilde{\phi}\nu_R\left(\phi^\dagger\phi\right) , \tag{8}$$

via the diagrams in Fig. 3. Solving the renormalization group equations we find that for $\Lambda \gtrsim 1$ TeV,

$$\mu_\nu \lesssim 8 \times 10^{-15}\mu_B\left(\frac{m_\nu}{1\text{ eV}}\right) , \tag{9}$$

in the absence of fine tuning.[10]

3. Majorana Neutrinos

We have seen above that the "naturalness" bounds on the magnetic moments of Dirac neutrinos are significantly stronger than present experimental limits. However, the analogous bounds for Majorana neutrinos are much weaker. The case of Majorana neutrinos is more subtle, due to the relative flavor symmetries of m_ν and μ_ν respectively. Majorana neutrinos cannot have diagonal magnetic moments, but are permitted non-zero transition moments. The transition magnetic moment $[\mu_\nu]_{\alpha\beta}$ is antisymmetric in the flavor indices $\{\alpha, \beta\}$, while the mass terms $[m_\nu]_{\alpha\beta}$ are symmetric. These different flavor symmetries play an important role in our limits, and are the origin of the difference between the magnetic moment constraints for Dirac and Majorana neutrinos.

As before, we write down the most general set of operators that can give rise to neutrino magnetic moment and mass terms, while respecting the SM gauge group. In the case of Majorana neutrinos, the lowest order contribution to the neutrino mass arises from the usual five dimensional operator containing Higgs and left-handed lepton doublet fields:

$$[O_M^{5D}]_{\alpha\beta} = \left(\overline{L_\alpha^c}\epsilon\phi\right)\left(\phi^T\epsilon L_\beta\right), \tag{10}$$

where $\epsilon = -i\tau_2$, $\overline{L^c} = L^T C$, C denotes charge conjugation, and α, β are flavor indices. The lowest order contribution to the neutrino magnetic moment arises from the following dimension seven operators,

$$[O_B]_{\alpha\beta} = g'\left(\overline{L^c}_\alpha\epsilon\phi\right)\sigma^{\mu\nu}\left(\phi^T\epsilon L_\beta\right)B_{\mu\nu}, \tag{11}$$

$$[O_W]_{\alpha\beta} = g\left(\overline{L_\alpha^c}\epsilon\phi\right)\sigma^{\mu\nu}\left(\phi^T\epsilon\tau^a L_\beta\right)W_{\mu\nu}^a, \tag{12}$$

and we also define a 7D mass operator as

$$[O_M^{7D}]_{\alpha\beta} = \left(\overline{L_\alpha^c}\epsilon\phi\right)\left(\phi^T\epsilon L_\beta\right)\left(\phi^\dagger\phi\right). \tag{13}$$

Operators O_M^{5D} and O_M^{7D} are flavor symmetric, while O_B is antisymmetric. The operator O_W is the most general 7D operator involving $W_{\mu\nu}^a$. However, as it is neither flavor symmetric nor antisymmetric it is useful to express it in terms of operators with explicit flavor symmetry, $O_W^\pm$, which we define as

$$[O_W^\pm]_{\alpha\beta} = \frac{1}{2}\left\{[O_W]_{\alpha\beta} \pm [O_W]_{\beta\alpha}\right\}. \tag{14}$$

Our effective Lagrangian is therefore

$$\mathcal{L} = \frac{C_M^{5D}}{\Lambda}O_M^{5D} + \frac{C_M^{7D}}{\Lambda^3}O_M^{7D} + \frac{C_B}{\Lambda^3}O_B + \frac{C_W^+}{\Lambda^3}O_W^+ + \frac{C_W^-}{\Lambda^3}O_W^- + \cdots \,. \tag{15}$$

After spontaneous symmetry breaking, the flavor antisymmetric operators O_B and O_W^- generate a contribution to the magnetic moment interaction $\frac{1}{2}[\mu_\nu]_{\alpha\beta}\,\overline{\nu^c}_\alpha\sigma^{\mu\nu}\nu_\beta F_{\mu\nu}$, given by

$$\frac{[\mu_\nu]_{\alpha\beta}}{\mu_B} = \frac{2m_e v^2}{\Lambda^3}\left([C_B(M_W)]_{\alpha\beta} + \left[C_W^-(M_W)\right]_{\alpha\beta}\right), \tag{16}$$

where the Higgs vacuum expectation value is $\langle\phi^T\rangle = (0, v/\sqrt{2})$. Similarly, the operators O_M^{5D} and O_M^{7D} generate contributions to the Majorana neutrino mass terms, $\frac{1}{2}[m_\nu]_{\alpha\beta}\overline{\nu^c}_\alpha\nu_\beta$, given by

$$\frac{1}{2}[m_\nu]_{\alpha\beta} = \frac{v^2}{2\Lambda}\left[C_M^{5D}(M_W)\right] + \frac{v^4}{4\Lambda^3}\left[C_M^{7D}(M_W)\right]. \tag{17}$$

Below, we outline the radiative corrections to the neutrino mass operators (O_M^{5D} and O_M^{7D}) generated by the magnetic moment operators O_W^- and O_B. This allows us to determine constraints on the size of the magnetic moment in terms of the neutrino mass, using Eqs.(16) and (17). Our results are summarized in Table 1 below, where we have defined $R_{\alpha\beta} = m_\tau^2/|m_\alpha^2 - m_\beta^2|$, with m_α being the masses of charged lepton masses. Numerically, one has $R_{\tau e} \simeq R_{\tau\mu} \simeq 1$ and $R_{\mu e} \simeq 283$.

Table 1. Summary of constraints on the magnitude of the magnetic moment of Majorana neutrinos. The upper two lines correspond to a magnetic moment generated by the O_W^- operator, while the lower two lines correspond to the O_B operator.

i) 1-loop, 7D	$\mu^W_{\alpha\beta}$	$\leq 1\times 10^{-10}\mu_B \left(\frac{[m_\nu]_{\alpha\beta}}{1\text{ eV}}\right) \ln^{-1}\frac{\Lambda^2}{M_W^2} R_{\alpha\beta}$
ii) 2-loop, 5D	$\mu^W_{\alpha\beta}$	$\leq 1\times 10^{-9}\mu_B \left(\frac{[m_\nu]_{\alpha\beta}}{1\text{ eV}}\right) \left(\frac{1\text{ TeV}}{\Lambda}\right)^2 R_{\alpha\beta}$
iii) 2-loop, 7D	$\mu^B_{\alpha\beta}$	$\leq 1\times 10^{-7}\mu_B \left(\frac{[m_\nu]_{\alpha\beta}}{1\text{ eV}}\right) \ln^{-1}\frac{\Lambda^2}{M_W^2} R_{\alpha\beta}$
iv) 2-loop, 5D	$\mu^B_{\alpha\beta}$	$\leq 4\times 10^{-9}\mu_B \left(\frac{[m_\nu]_{\alpha\beta}}{1\text{ eV}}\right) \left(\frac{1\text{ TeV}}{\Lambda}\right)^2 R_{\alpha\beta}$

3.1. *SU(2) gauge boson*

3.1.1. *7D mass term — O_W*

As the operator O_W^- is flavor antisymmetric, it must be multiplied by another flavor antisymmetric contribution in order to produce a flavor symmetric mass term. This can be accomplished through insertion of Yukawa couplings in the diagram shown in Fig. 4.[11] This diagram provides a logarithmically divergent contribution to the 7D mass term, given by[11]

$$\left[C_M^{7D}(M_W)\right]_{\alpha\beta} \simeq \frac{3g^2}{16\pi^2}\frac{m_\alpha^2 - m_\beta^2}{v^2}\ln\frac{\Lambda^2}{M_W^2}\left[C_W^-(\Lambda)\right]_{\alpha\beta}, \tag{18}$$

where m_α are the charged lepton masses, and the exact coefficient has been computed using dimensional regularization. Using this result, together with Eqs. (16) and (17), leads to bound (i) in Table 1.

3.1.2. *5D mass term — O_W*

The neutrino magnetic moment operator O_W^- will also contribute to the 5D mass operator via two-loop diagrams, as shown in Fig. 5.[12] As with the diagrams in Fig. 4, we require two Yukawa insertions in order to obtain a flavor symmetric result. Using dimensional analysis, we estimate[12]

$$\left[C_M^{5D}(\Lambda)\right]_{\alpha\beta} \simeq \frac{g^2}{(16\pi^2)^2}\frac{m_\alpha^2 - m_\beta^2}{v^2}\left[C_W^-(\Lambda)\right]_{\alpha\beta}. \tag{19}$$

This leads to bound (ii) in Table 1. Compared to 1-loop (7D) case of Eq. (18), the 2-loop (5D) mass contribution is suppressed by a factor of $1/16\pi^2$ arising from the additional loop, but enhanced by a factor of Λ^2/v^2 arising from the lower operator dimension. Thus, as we increase the new physics scale, Λ, this two-loop constraint rapidly becomes more restrictive. The "crossover" scale between the two effects occurs at ~ 10 TeV.

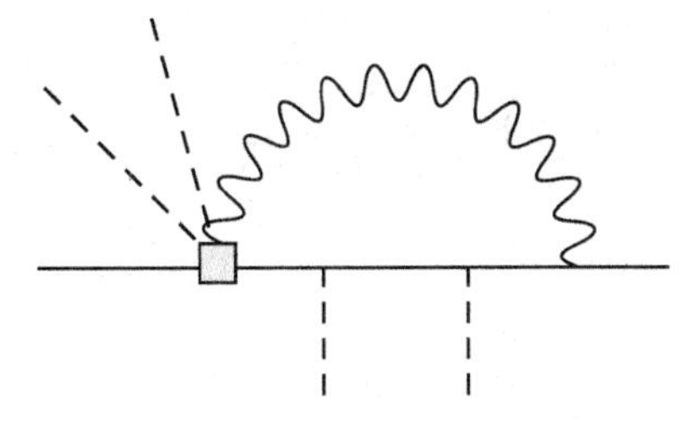

Fig. 4. Contribution of $O_{\tilde{W}}$ to the 7D neutrino mass operator.

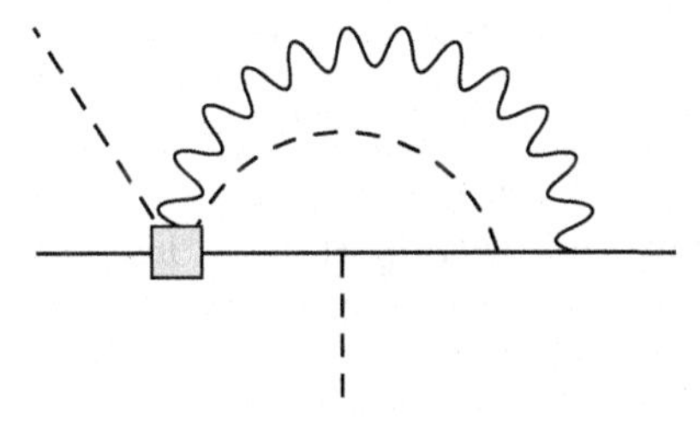

Fig. 5. Representative contribution of $O_{\tilde{W}}$ to the 5D neutrino mass operator.

3.2. *Hypercharge gauge boson*

3.2.1. *7D mass term — O_B*

If we insert O_B in the diagram in Fig. 4, the contribution vanishes, due to the $SU(2)$ structure of the graph. Therefore, to obtain a non-zero contribution to O_M^{7D} from O_B we require the presence of some non-trivial $SU(2)$ structure. This can arise, for instance, from a virtual W boson loop as in Fig. 6.[11] This mechanism gives the leading contribution of the operator O_B to the 7D mass term. The O_B and O_W contributions to the 7D mass term are thus related by

$$\frac{(\delta m_\nu)^B}{(\delta m_\nu)^W} \approx \frac{\alpha}{4\pi} \frac{1}{\cos^2 \theta_W}, \tag{20}$$

where θ_W is the weak mixing angle and where the factor on the RHS is due to the additional $SU(2)_L$ boson loop. This additional loop suppression for the O_B contribution results in a significantly weaker neutrino magnetic moment constraint than that obtained above for $O_{\tilde{W}}$. The corresponding limit is shown as bound (iii) in Table 1.

3.2.2. *5D mass term — O_B*

However, the leading contribution of O_B to the 5D mass term arises from the same 2-loop diagrams (Fig. 5) that we discussed in connection with the $O_{\tilde{W}}$ operator. Therefore, the contribution to the 5D mass term is the same as that for O_W, except

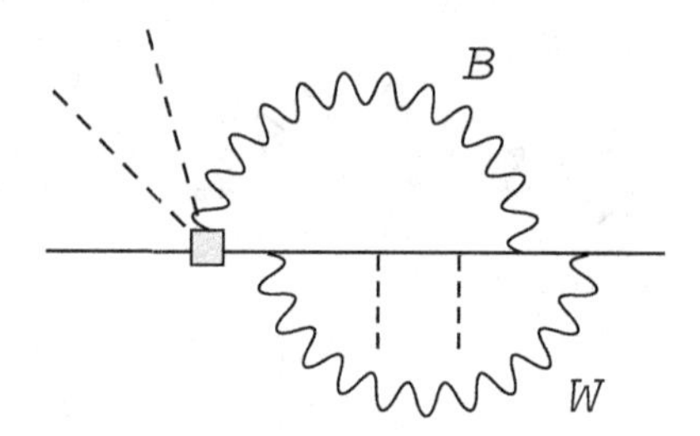

Fig. 6. Representative contribution of O_B to the 7D neutrino mass operator at two loop order.

for a factor of $(g'/g)^2 = \tan^2\theta_W$. We thus obtain[12]

$$\left[C_M^{5D}(\Lambda)\right]_{\alpha\beta} \simeq \frac{g'^2}{(16\pi^2)^2}\frac{m_\alpha^2 - m_\beta^2}{v^2}\left[C_B(\Lambda)\right]_{\alpha\beta} \quad , \tag{21}$$

corresponding to bound (iv) in Table. 1. Importantly, this is the strongest constraint on the O_B contribution to the neutrino magnetic moment for any value of Λ, and the most general of our bounds on $\mu_\nu^{\text{Majorana}}$.[12]

3.3. *Comparison with experimental limits*

The best laboratory limit on μ_ν, obtained from scattering of low energy reactor neutrinos is, "μ_e" $< 0.7 \times 10^{-10}\mu_B$.[3] Note that this limit applies to both $\mu_{\tau e}$ and $\mu_{\mu e}$, as the flavor of the scattered neutrino is not detected in the experiment. Taking the neutrino mass to be $m_\nu \lesssim 0.3$ eV (as implied by cosmological observations, e.g. Ref. 13), bound (iv) in Table. 1 gives

$$\begin{aligned} \mu_{\tau\mu}, \mu_{\tau e} &\lesssim 10^{-9}\,[\Lambda(\text{TeV})]^{-2} \\ \mu_{\mu e} &\lesssim 3 \times 10^{-7}\,[\Lambda(\text{TeV})]^{-2} \,. \end{aligned} \tag{22}$$

For Majorana neutrinos we thus conclude that if $\mu_{\mu e}$ is dominant over the other flavor elements, an experimental discovery near the present limits (e.g., at $\mu \sim 10^{-11}\mu_B$) would imply $\Lambda \lesssim 100$ TeV. However, this would become $\Lambda \lesssim 10$ TeV in any model in which all element of $\mu_{\alpha\beta}$ have similar size.

4. Conclusions

We have discussed radiative corrections to the neutrino mass arising from a neutrino magnetic moment coupling. Expressing the magnetic moment in terms of effective operators in a model independent fashion required constructing operators containing the $SU(2)_L$ and hypercharge gauge bosons, O_W and O_B respectively, rather than working directly with the electromagnetic gauge boson. We then calculated μ_ν naturalness bounds arising from the leading order contributions to neutrino mass term, for both Dirac and Majorana neutrinos. For Dirac neutrinos we found

$$\mu_\nu^{\text{Dirac}} \lesssim 3 \times 10^{-15}\mu_B\left(\frac{m_\nu}{1\ \text{eV}}\right)\left(\frac{1\ \text{TeV}}{\Lambda}\right)^2 , \tag{23}$$

while the most general naturalness bound on the size of the Majorana neutrino magnetic moment is

$$\mu_{\alpha\beta}^{\text{Majorana}} \leq 4 \times 10^{-9} \mu_B \left(\frac{[m_\nu]_{\alpha\beta}}{1 \text{ eV}} \right) \left(\frac{1 \text{ TeV}}{\Lambda} \right)^2 \left| \frac{m_\tau^2}{m_\alpha^2 - m_\beta^2} \right| . \tag{24}$$

These limits can only be evaded in the presence of fine tuning.

The limit on the the magnetic moments of Dirac neutrinos is thus considerably more stringent than for Majorana neutrinos. This is due to the different flavor symmetries involved, since in the Majorana case we require the insertion of Yukawa couplings to convert a flavor antisymmetric (magnetic moment) operator into a flavor symmetric (mass) operator. Our results implies that an experimental discovery of a magnetic moment near the present limits would signify (i) neutrinos are Majorana fermions and (ii) new lepton number violating physics responsible for the generation of μ_ν arises at a scale Λ which is well below the see-saw scale.

Acknowledgments

This article is based upon the results of Ref. 10 and Ref. 12. NFB thanks Vincenzo Cirigliano, Mikhail Gorchtein, Michael Ramsey-Musolf, Petr Vogel, Peng Wang and Mark Wise for an enjoyable and productive collaboration.

References

1. W. J. Marciano and A. I. Sanda, *Phys. Lett. B* **67**, 303 (1977); B. W. Lee and R. E. Shrock, *Phys. Rev. D* **16**, 1444 (1977); K. Fujikawa and R. Shrock, *Phys. Rev. Lett.* **45**, 963 (1980).
2. J. F. Beacom and P. Vogel, *Phys. Rev. Lett.* **83**, 5222 (1999); D. W. Liu *et al.*, *Phys. Rev. Lett.* **93**, 021802 (2004).
3. H. T. Wong *et al.* [TEXONO Collaboration], *Phys. Rev. D* **75**, 012001 (2007); B. Xin *et al.* [TEXONO Collaboration], *Phys. Rev. D* **72**, 012006 (2005); Z. Daraktchieva *et al.* [MUNU Collaboration], *Phys. Lett. B* **615**, 153 (2005).
4. G.G. Raffelt, *Phys. Rep.* **320**, 319 (1999).
5. M. B. Voloshin, *Sov. J. Nucl. Phys.* **48**, 512 (1988). For a specific implementation, see R. Barbieri and R. N. Mohapatra, *Phys. Lett. B* **218**, 225 (1989).
6. H. Georgi and L. Randall, *Phys. Lett. B* **244**, 196 (1990).
7. W. Grimus and H. Neufeld, *Nucl. Phys. B* **351**, 115 (1991).
8. K. S. Babu and R. N. Mohapatra, *Phys. Rev. Lett.* **64**, 1705 (1990).
9. S. M. Barr, E. M. Freire and A. Zee, *Phys. Rev. Lett.* **65**, 2626 (1990).
10. N. F. Bell, V. Cirigliano, M. J. Ramsey-Musolf, P. Vogel and M. B. Wise, *Phys. Rev. Lett.* **95**, 151802 (2005).
11. S. Davidson, M. Gorbahn and A. Santamaria, *Phys. Lett. B* **626**, 151 (2005).
12. N. F. Bell, M. Gorchtein, M. J. Ramsey-Musolf, P. Vogel and P. Wang, *Phys. Lett. B* **642**, 377 (2006).
13. D. N. Spergel *et al.*, astro-ph/0603449.

ANIONS AND ANOMALIES

M. BAWIN

University of Liège
Institut de Physique B5,
Sart Tilman, 4000 Liège 1, Belgium

SIDNEY A. COON*

Office of Nuclear Physics
SC-26, Germantown Building
U. S. Department of Energy
1000 Independence Avenue, SW
Washington, D.C. 20585-1290

BARRY R. HOLSTEIN

Department of Physics-LGRT
University of Massachusetts
Amherst, MA 01003

We analyze the recent claim that experimental measurements of binding energies of dipole-bound anions can be understood in terms of a quantum mechanical anomaly. The discrepancy between the experimental critical dipole moments and that predicted by the anisotropic inverse square potential of a static dipole precludes such an explanation. As has long been known, in the physical problem one must include rotational structure so that the long distance behavior changes from $1/r^2$ to $1/r^4$. In a simple model this can be shown to lead to a modification of the critical dipole moment of 20% or so, bringing it into agreement with experiment. This, together with the fact that inclusion of finite size effects does not change the critical dipole moment of the static point dipole, strongly suggests that the quantum mechanical anomaly interpretation of the formation of dipole-bound anions cannot be correct.

1. Introduction

Singular potentials in quantum mechanics have been studied by a number of authors, including Mott and Massey,[1] Titchmarsh,[2] Case,[3] Landau and Lifshitz,[4] Meetz,[5] Jackiw[6] and others.[7] The Schrödinger equation with a singular potential does not admit physically meaningful solutions with the usual boundary conditions. The attractive singular potential $1/r^2$ is especially interesting because it is on the boundary which separates a regular power law potential like $1/r$ from truly singular potentials like $1/r^3$ (*e.g.*, from one-pion-exchange models of the nucleon-nucleon

---

*Speaker.

interaction) or the familiar $1/r^4$ polarization potential of electrostatics. The latter potentials are singular for all values of of the attractive strength ("coupling") but the weakly attractive $1/r^2$ potential is regular and becomes singular only for couplings greater than a certain critical value. About thirty years ago, high energy scattering experiments at SLAC focused attention on an approximate space-time symmetry called scale or dilation invariance and it was soon realized that only a Hamiltonian with a $1/r^2$ potential is scale invariant in any number of dimensions.[6,8] Nöther's theorem, which guarantees the correspondence between symmetries and conservation laws, then implies a physical consequence if the symmetry is broken.

The subject of the "not always singular" $1/r^2$ potential was revived recently by Camblong et al. in the context of regularization methods[9] which break the scale invariance of a quantum mechanical system. The problem of an electron in the electric dipole field of a static polar molecule has long been solved;[10] above a critical moment the electron is captured and the resulting bound state is called a dipole-bound anion (hereafter shortened to simply anion as we do not consider valence anions). Camblong et al. suggest that the observed formation of anions is a realization in quantum mechanics of a physical consequence called anomalous or quantum symmetry breaking; in this case the breaking of scale invariance by a particular regularization method.[11] On this basis, two of us (BRH and SAC) have even written a didactic article on this subject.[12] However, while the mathematics of the model which interprets these anion experiments as an example of quantum mechanical symmetry breaking is certainly correct, we now recognize that the physical connection proposed in Ref. 11 *is not*[13] and the purpose of the present note is to validate this assertion. Before doing so, let us make some more comments about the physical interest of this problem. Old[14–16] and recent,[18,17,19] calculations based upon conventional quantum mechanics with non-singular potentials provide a good description of the experimentally observed properties of anions.It is nevertheless of great interest in physics to study, as done by Camblong et al.[9] whether given properties of a system can be ascribed to a (possibly broken) fundamental symmetry, and it is accordingly important to check whether the scheme proposed in Ref. 11 agrees with experimental data.Camblong et al. do not make such a comparison with data but simply assert that their critical dipole moment is "of a similar value for a large number of molecules".[11]

In section 2 we review the mathematics of the $1/r^2$ potential and its relationship to the anion experiments. Then, in section 3 we show, within a simple dipole potential model, how the experiments *can* be understood in terms of one-electron models of anion formation which include the effects of rotational structure. The more inclusive one-electron models used to fit the data do of course include the rotational degrees of freedom of the molecule. What we claim is that the latter degree of freedom is essential to our understanding of anion binding energies because it provides the leading contribution to the long range part of the molecular potential (see Ref. 13 and references therein). This important feature was of course discussed in previous work on the subject[14–17] but, we feel, deserves a simple analytical illustration after the interpretation of Camblong et al.,[11] which does not include the effects of rotation. We summarize our findings in a brief concluding section.

2. Regularization of the $1/r^2$ Potential

In this section we present a succinct review of the arguments that lead to a critical binding energy. We consider the one dimensional Hamiltonian

$$H = -\frac{1}{2m}\frac{d^2}{dx^2} + V(x) \tag{1}$$

with

$$V(x) = \begin{cases} -\frac{\beta}{x^2} & x \geq 0 \\ \infty & x < 0 \end{cases} \tag{2}$$

so that positive β corresponds to an attractive potential. If $0 \leq 2m\beta \leq 1/4$, the potential is is weakly attractive, there exist no bound states, and the scattering states form a continuum with energies extending from zero to positive infinity, just as expected from the scale invariance of the Hamiltonian. For a stronger attractive potential (which is now singular) a problem arises, since the Schrödinger equation is invariant under the scale transformation

$$x \to \lambda x \quad \text{and} \quad k \to k/\lambda$$

where the energy is given in terms of k by $E = k^2/2m$. The problem is clear. Suppose that one bound state exists, as signified by the existence of a negative energy solution to the Schrödinger equation. By scale invariance, either the bound state is unique with energy of negative infinity, or (as is the case with the inverse square potential) there must also exist an entire series of such bound state solutions with energies extending to negative infinity. Either alternative, including the latter one, which holds here, is physically unacceptable.

The solution to this problem is regularization.[20] Defining $\gamma = \sqrt{2m\beta - \frac{1}{4}}$ and $E = -\kappa^2/2m < 0$ the negative energy solutions are easily seen to be modified Bessel functions of order $i\gamma$

$$\psi(x) = I_{i\gamma}(\kappa x),\ K_{i\gamma}(\kappa x). \tag{3}$$

Ordinarily, if we imagine a wall at $x = 0$, we would impose the boundary condition $\psi(x = 0) = 0$. However, in our case neither function is well defined at the origin, and in addition, the function $I_{i\gamma}(\kappa x)$ diverges for large x and is not square integrable. In order to have a well-defined bound state solution we need to modify the parameters of our problem. There are a number of ways by which this can be achieved but perhaps the simplest is move the infinite "wall" from the origin to $x = a$, where $a > 0$ is a small but arbitrary displacement. The boundary condition becomes $\psi(x = a) = 0$ and the system is now well defined. A solution is possible because the introduction of the parameter a has broken the scale invariance and the resulting bound state energy is found to be[20]

$$E_n = -\frac{4}{a^2}\exp 2[\arg\Gamma(1+i\gamma) - n\pi]/\gamma \quad n = n_0, n_0+1, \ldots \tag{4}$$

In the case of positive energy, we can also consider scattering solutions and determine the scattering phase shift[12]

$$\tan\delta = \frac{\pi - 2\log\frac{k}{\kappa_{min}}}{\pi + 2\log\frac{k}{\kappa_{min}}} \tag{5}$$

where $\kappa_{min} = \sqrt{-2mE_{n_0}}$. Now since a is arbitrary, it is impossible to *predict* κ_{min}. However, once the ground state energy is determined the scattering is uniquely defined via Eq. (5).

The key to this problem is to take the limit as the boundary a approaches the origin. In order that that the physics be independent of the regularization parameter a, the index γ must also vanish, and in this limit—$2m\beta = 1/4$—there can exist a single bound state, characterized by n_0. Any other value of the coupling β cannot support a bound state.

So far, this is a simple mathematical model. A possible connection with experimental physics can be made by realizing that a simple system in three dimensions consisting of a point dipole and point charge reduces to a system of coupled differential equations, one of which is similar to the one-dimensional case above. Specifically, we note that the potential energy of such a system is given by

$$V(\vec{r}) = \frac{q\vec{p}\cdot\vec{r}}{4\pi r^3} \tag{6}$$

where $\vec{p}$ is the electric dipole moment. The resulting Schrödinger equation

$$-\nabla^2\psi + \frac{\sigma\cos\theta}{r^2}\psi = k^2\psi \tag{7}$$

is separable, yielding the coupled equations

$$\begin{aligned} -\frac{d^2u(r)}{dr^2} - \frac{c}{r^2}u(r) &= k^2u(r) \\ -\sigma\cos\theta Y(\theta) - \hat{L}^2Y(\theta) &= cY(\theta) \end{aligned} \tag{8}$$

where c is the separation constant and $\sigma = 2mqp/4\pi$. The angular equation is an eigenvalue equation which may be solved numerically.[10–12] Using an angular momentum basis it is found that there exists a critical value

$$\sigma_{crit} = 1.2786...$$

corresponding to $c = 1/4$. There thus exists a critical dipole moment

$$p_{crit} = \frac{4\pi}{mq} \times 1.2786$$

such that for dipole values smaller than this no binding is possible, and the regularized potential supports a single bound state. In this context it was noted by

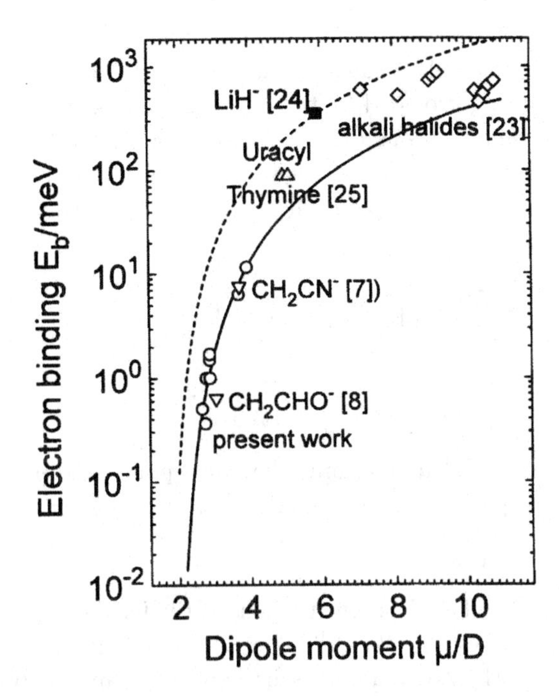

Fig. 1. Shown are data reported in Ref. 18 on experimental anion binding energies. The solid and dotted curves are the result of one-electron models with a pseudopotential between the excess electron and the molecule. Note that there appears to exist a limiting value $p_{crit} \simeq 2D$ to both data and the model.

Camblong et al.[11] that experiments on anions—electrons bound to molecular systems with a permanent dipole moment—show just such a critical moment of a similar value and on this basis they concluded that anion formation was indeed an example of an anomaly in the context of quantum mechanics[11]—*cf.* Figure 1.

2.1. *An alternate picture*

A more detailed analysis, however, leads us to question this claim. The critical value of the electric dipole moment in the point dipole model is $p_{crit} = 1.625 \times 10^{-18}$ esu cm $\equiv 1.625D$ where the Debye D is a characteristic scale for molecular dipole moments and is defined to be exactly $1D = 10^{-18}$ esu cm. Comparing this value with the experimental value for anion formation suggested by Figure 1

$$p_{exp} \simeq 2D$$

we see that the experimental number is about twenty percent larger than that predicted by the anomaly. Now in Ref. 12 this discrepancy was noticed but at the time we attributed the difference to the feature that the physical dipole to which the electron is bound has nonzero size, meaning that in addition to the scale invariance there exists an *explicit* symmetry breaking term. An example of this phenomenon in quantum field theory can be seen in $\pi^0 \to 2\gamma$ decay wherein the chiral anomaly predicts a rate for this process which is modified (increased about five per cent) by the chiral symmetry breaking provided by the existence of nonzero quark mass.[21]

However, we no longer make this assertion, since an exact solution for an electron bound to a physical dipole of arbitrary size can be obtained for the case that the dipole is constructed from a pair of point charges.[10] In this case the Schrödinger equation becomes

$$-\frac{1}{2m}\nabla^2\psi + \frac{qQ}{4\pi}\left(\frac{1}{r_+} - \frac{1}{r_-}\right)\psi = \frac{k^2}{2m}\psi \tag{9}$$

where

$$r_\pm = \sqrt{r^2 + \frac{l^2}{4} \pm rl\cos\theta}\ ,$$

l is the length of the extended dipole, and the corresponding dipole moment is $p = Ql$. The equation is separable in spheroidal coordinates

$$\xi = \frac{1}{2d}(r_+ + r_-), \quad \zeta = \frac{1}{2}(r_+ - r_-), \quad \phi$$

where ϕ denotes the azimuthal angle about the axis of the fixed charges, and can be exactly solved as shown by Levy-LeBlond.[10] The system is found to have a critical value *identical* to that for the point dipole—the explicit symmetry breaking does *not* affect the critical moment. This feature of the model for anion formation was emphasized by Camblong et al. in Ref. 11 as evidence for the robustness of their anomaly interpretation of anion formation, and this was unfortunately overlooked in Ref. 12. In any event, we now suggest that the twenty percent discrepancy (not explicated in Ref. 11) between the experimental and theoretical values of the critical moment for anion binding energies does *not* not support interpretation in terms of a quantum mechanical anomaly and subsequent symmetry breaking by finite size effects.

3. An Explanation

That the experimental results shown in Figure 1 are *not* connected with a quantum mechanical anomaly is the primary conclusion of our paper. However, if this is not the explanation, it behooves us to provide an alternative picture. In fact, it has been shown that anion binding energies *can* be understood via a one-electron pseudopotential model[17–19] or even multi-electron *ab initio* studies (reviewed in Ref. 17, for example), without explicit reference to any renormalization scheme or to a quantum mechanical anomaly. These calculations provide quantitative predictions of the energy of an electron loosely bound not only by the dipole moment of the molecule, and in addition include a possible quadrupole moment, induced dipole moment and polarization force. Our simple picture of anion binding energies, an alternative to the anomaly interpretation we now reject, seeks to understand the basic experimental results with the aid of a simple electron-dipole model, but does not attempt to fit data, in contrast to the contemporary one-electron pseudopotential models.

One of us (MB) recently argued[13] that the crucial ingredient of the molecular calculations was the description of neutral molecular cores as *rotating extended dipoles*[15] rather than point static dipoles as in Ref. 11. The current detailed molecular calculations motivated by experiment use Clary's adiabatic theory[16] for the eigenstates of the bound electron. In this picture, the electron tends to minimize its interaction with the molecular dipole (see Eq.(6)) and thus the electron wavefunction adiabatically follows the rotational motion of the molecule. As advocated by Clary, the bound eigenstates are calculated by assuming that the electron moves on the lowest rotationally adiabatic potential obtained by diagonalization of the full Hamiltonian. This Hamiltonian contains the rotational energy of the neutral molecule (thereby introducing the rotor constant B), as well as the relative angular momentum of the electron and the electron-molecule interaction. This latter interaction can be the schematic point dipole of Eq. (6), the physical dipole of Eq. (9) or the more quantitative interactions of Refs. 17–19. In all cases, however, this procedure leads to a Schrödinger equation with an effective potential whose asymptotic behavior is $\sim 1/r^4$ $(B \neq 0)$ rather than $\sim 1/r^2$ $(B = 0)$. Within a good approximation, the spectrum of this effective potential is like that of its static parent: all rotational levels are simply shifted down to lower energies.

In addition, the effective potential from the extended dipole of Eq. (9) is not singular but is regular near the origin. This is not the case for the point dipole of Eq. (6); merely rotating a singular potential does not change its singularity structure. Therefore, it is not surprising that Clary's procedure yields[17] a critical value of $\sim 1.65D$ for dipole-binding to a rotating point dipole (at zero total angular momentum), correct to within 2 per cent of $p_{crit} = 1.625D$ of the previous section. As shown in Ref. 17 and to be illustrated in our simple model, this result may be significantly changed for an extended rotating dipole.

To summarize, it was shown in Ref. 13 that Clary's adiabatic theory[16] for electron eigenstates in the field of dipolar molecules leads to a Schrödinger equation with an effective potential whose behavior at the origin is regularized by the extended nature of the dipole and whose asymptotic behavior is $1/r^4$ rather than $1/r^2$ due to its rotational motion. Those two features of the effective potential:

i) regular near the origin for an extended dipole, and
ii) asymptotically $\sim B/r^4$ because of rotation,

together account for the salient features of the dipole bound anion, *i.e.*, the existence of one or two weakly bound states above some critical value of the molecular dipole moment. No renormalization scheme is needed to obtain these features. This demonstration[13] corroborates detailed calculations[17–19] but it depends only on upper bounds for the number of bound states of a Schrödinger particle moving in an attractive spherically symmetric potential.

We go on to reproduce here in an extremely simple model another feature of Figure 1—that data and the theoretical curves *both* indicate a p_{crit} greater than $\sim 1.6D$. The theoretical curves of Figure 1 include all the complexities of the realistic

Hamiltonians but it is a striking observation (*cf.* Figure 3 in Ref. 17) that p_{crit} is about $2D$ even for the pure extended (and rotating) dipole case when the other features (polarizability potentials etc.) of the realistic problem are turned off. The only possible conclusion is that the rotor constant B drastically changes the value of p_{crit} from the value predicted by the "anomaly" interpretation.

We now proceed to demonstrate this assertion by studying analytically in a simple model how the critical value changes with nonzero B from the well known one for $B = 0$ (no rotation). In order to do so, we follow a didactic approach and take as our starting point the three dimensional Schrödinger equation describing a particle of mass m in the potential $V(r)$:

$$V(r) = \begin{cases} \frac{-\alpha}{r^2} & (r < b) \\ \frac{-\alpha b^2}{r^4} & (r > b) \end{cases} . \tag{10}$$

This effective potential describes the interaction with a point dipole at short distance but includes the rotational structure of the system at large distance. Let us emphasize that we purposely use such a simple model in order to isolate the crucial features of the molecular interaction for the discussion at hand. Detailed numerical calculations with more elaborate potentials which do describe the data can be found in Refs. 14, 18 and 19.

As usual, we define:

$$\gamma = (2m\alpha - 1/4)^{1/2} \tag{11}$$

and we regularize the potential by the condition ($a << b$ but otherwise arbitrary)

$$\psi(r = a) = 0 \tag{12}$$

Then the critical value would be given by $2m\alpha = 1/4$ if b were infinite and the potential were inverse square throughout all space. In order to study the dependence of the critical value upon b^2, we need only find the *zero energy* solution $\psi_0(r)$ to the radial equation, since it is known that the number of nodes of $\psi_0(r)$ is the number of bound states of the system.[22] Requiring the reduced wavefunction $\psi_0(r)$ and its derivative to match at $r = b$ leads to:

$$\psi_0(r) = \begin{cases} (r/a)^{1/2} \sin(\gamma \log(\frac{r}{a})) & r < b \\ (b/a)^{1/2}(\frac{r}{b}) \frac{\sin(\gamma \log \frac{b}{a}) \sin(\frac{(2m\alpha)^{1/2} b}{r} + \beta)}{\sin((2m\alpha)^{1/2} + \beta)} & r > b \end{cases} \tag{13}$$

with

$$\beta = \text{arccot}[\frac{1/2 - \gamma \cot \gamma \log(b/a)}{(2m\alpha)^{1/2}}] - (2m\alpha)^{1/2} . \tag{14}$$

Now from Eq. (13), the wavefunction $\psi_0(r)$ is proportional to $\sin \beta$ in the limit as $r \to \infty$. Thus the first zero of $\psi_0(r)$ will occur for $\beta = 0$. Figure 2 shows the form of β as a function of $\lambda = (2m\alpha)^{1/2}$ for $b/a = 1000$.

We see then that β vanishes for $\lambda \sim 0.61$. That this is a critical value can be seen clearly in Figure 3, which shows $\psi_0(r)$ for $\lambda = 0.595$. Obviously, there is no zero and hence no bound state.

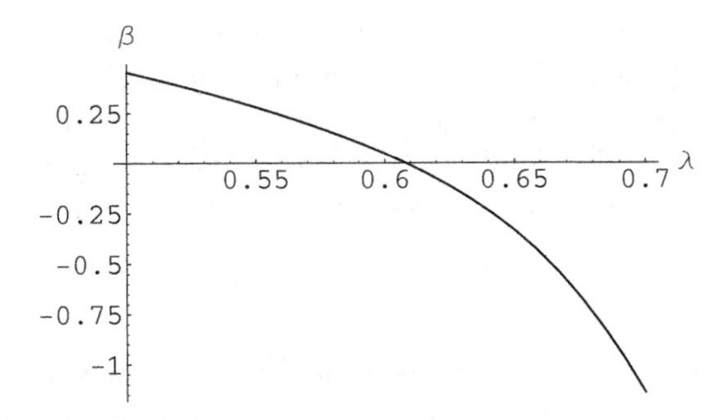

Fig. 2. β as a function of λ for $b/a = 1000$.

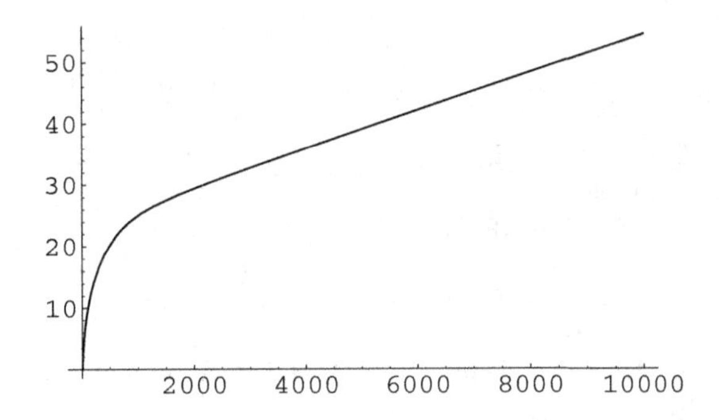

Fig. 3. $\psi_0(r)$ as a function of r (in units of b) for $\lambda = 0.595$ and $b/a = 1000$.

We conclude there exists a modification in the critical coupling (from $\lambda^2 = 1/4$) due to the change in the long range part of the interaction. We have checked that the critical coupling approaches $\lambda = 0.50$ as b/a becomes very large, as expected. It is amusing that with the chosen value of b/a there is an increase of about 20 percent in the value of the critical coupling. The same feature occurs in models of molecular physics which have static dipole moments versus rotating physical dipole moments. Of course, anything more than this crude sort of speculation is not warranted given the roughness of our model. Nevertheless, we believe that this simple model (which even artificially regularizes the point dipole at short distances Eq. (12)) illustrates the salient features of the molecular physics problem.

4. Conclusion

Above we have studied the physics associated with bound states of physical molecules and electrons. In section 2 we reviewed the claim that the existence of a critical dipole moment is the simple manifestation of anomalous or quantum mechanical symmetry breaking, as proposed by Camblong et al. However, there exists a twenty percent discrepancy between the critical coupling predicted in such a picture and that found experimentally. We argued that the physical size of the molecule cannot account for such a shift, casting serious doubt on an anomaly explanation for the observed critical moment. In section 3, we used a simple model in order to isolate the essential physics which is involved here and showed that it is easy to understand the experimental findings without invoking quantum mechanical symmetry breaking, provided that the rotational structure is taken into account.

Acknowledgments

This work of M.B. is supported by the Belgian National Fund for Scientific Research, while that of BRH is supported by the U.S. National Science Foundation under award PHY05-53304.

References

1. N.F. Mott and H.S.W. Massey, *The Theory of Atomic Collisions*, Clarendon, Oxford (1933).
2. E.C. Titchmarsh, *Eigenfunction Expansions Associated with Second-Order Differential Equations*, Clarendon, Oxford (1946).
3. K.M. Case, *Phys. Rev.* **80**, 797 (1950).
4. L.D. Landau and E.M. Lifshitz, *Quantum Mechanics: Non-Relativistic Theory*, Addison-Wesley, Reading, MA (1958).
5. K. Meetz, *Nuovo Cim.* **34**, 690 (1964).
6. R. Jackiw, in *M.A.B. Memorial Volume*, A. Ali and P. Hoodbhoy, eds. (World Scientific, Singapore, 1991) pp. 25-42.
7. For a review through 1969, see W. M Frank, D. J. Land and R. M. Spector, *Rev. Mod. Phys.* **43**, 36 (1971).
8. R. Jackiw, *Physics Today* **25**, 23-27 (January 1972).
9. H.E. Camblong, L.N. Epele, H. Fanchiotte, and C.A. Garcia Canal, *Ann. Phys. (NY)* **287**, 14 and 57 (2001).
10. J.M. Lévy-Leblond, *Phys. Rev.* **153**, 1 (1967); for an account of the derivation of the value of the crictical moment presented by E. Fermi and E. Teller twenty years earlier, see J.E. Turner, *Am. J. Phys.* **45**, 758 (1977).
11. H.E. Camblong, L.N. Epele, H. Fanchiotti, and C.A. Garcia Canal, *Phys. Rev. Lett.* **87**, 220402 (2001).
12. S.A. Coon and B.R. Holstein, *Am. J. Phys.* **70**, 513 (2002).
13. M. Bawin, *Phys. Rev.* **A70**, 022505 (2004).
14. O.H. Crawford and W.R. Garrett, *J. Chem. Phys.* **66**, 4968 (1977)
15. W.R.Garrett, *Phys. Rev.* **A3**, 961 (1971) and *Phys. Rev.* **A22**, 1769 (1980).
16. D.C. Clary, *J. Phys. Chem.* **92**, 3173 (1988).
17. C. Desfrançois, H. Abdoul-Carime and J.P. Schermann, *Int. J. Mod. Phys.* **B12**, 1339 (1996).
18. C. Desfrançois, H. Abdoul-Carime, N. Khelifa and J.P. Schermann, *Phys. Rev. Lett.* **73**,2436 (1994)
19. H.Abdoul-Carime and C. Desfrançois, *Eur.Phys.J.* **D2**, 149 (1998).
20. K.S. Gupta and S.G. Rajeev, *Phys. Rev.* **D48**, 5940 (1993)
21. J. Goity, A.M. Bernstein, and B.R. Holstein, *Phys. Rev.* **D66**, 070014 (2002); B. Ananthanarayan and B. Moussallam, *JHEP* **0205**, 052 (2002).
22. F. Brau and F. Calogero, *J. Phys.* **A36**, 12021 (2003).

PROPERTY VALUES

R. DELBOURGO*

School of Mathematics and Physics, University of Tasmania,
Hobart, Tasmania 7001, Australia
**E-mail: bob.delbourgo@utas.edu.au*

By ascribing a complex anticommuting variable ζ to each basic *property* of a field it is possible to describe all the fundamental particles as combinations of only five ζ and understand the occurrence of particle generations. An extension of space-time x to include property then specifies the 'where, when and what' of an event and allows for a generalized relativity where the gauge fields lie in the $x - \zeta$ sector and the Higgs fields in the $\zeta - \zeta$ sector.

Keywords: Anticommuting coordinates; field properties; unified models.

Preamble

First of all let me say why I am so pleased to be present at this FestSchrift in honour of Girish' and Bruce's retirements. When I arrived in Tasmania one of my first tasks was to make contact with physicists in other Australian institutions. Amongst these Melbourne University had high priority and I was very glad to welcome Bruce as one of the earliest visitors to Hobart. Since then we have interacted many times and I have relished my own visits to Melbourne to give occasional seminars and find out what Joshi and McKellar were up to. Let me wish them both a long and happy retirement and say how much I have appreciated their support and friendship throughout the last 30 years! Retirement has much to recommend it and they might care to view their future condition as changing from 'battery hen' to 'free-range chicken', since they are no longer obliged to feed on grants in order to lay eggs.

1. Introduction

As most of you will have already surmised, the title of my talk has nothing to do with real estate and perhaps sounds all the more mysterious for it. In fact property values have everything do with quantum fields and accurately reflect the contents of what follows. The motivation for this work is to be able to describe the 'what' as well as the 'where-when' of an event. To make an analogy with personality traits in humans, psychologists may characterise a person as optimistic/pessimistic, happy/sad, aggressive/submissive, etc. (A person will possess some combination or superposition over these trait states.) The same sort of characterisation applies to a quantum field and is usually underlined by attaching a label to each distinct field. Thus we speak of an electron/positron field, a neutron/antineutron field, a quark/antiquark field, etc. and draw them in Feynman diagrams say with solid or dashed lines and arrows, plus legends if necessary, to distinguish them from one another. An event is some local confluence between these labels with a possible interchange of traits, as specified by an interaction Lagrangian. Often we assume a symmetry group is operational which 'rotates' labels; this constrains the interchange of property and, if the symmetry is local, it can be gauged.

The basic idea I wish to put to you is that traits/labels are normally discrete: thus a field is either an electron or it is not; similarly for a proton, and so on. It makes sense to attach a separate coordinate to each such property and choose the coordinate to be anticommuting since its occupation number is either one or zero. (So a person can be pessimistic & sad & aggressive with a product of the three traits.) Furthermore if we make the coordinate ζ complex we can describe the converse property by simple conjugation. When applying this idea to quantum fields the question arises: how many coordinates are needed?

The fewer the better of course. From my investigations thus far[1,2] I have concluded that one can get a reasonable description of the known fundamental particle spectrum by using just five complex cordinates ζ^μ; $\mu = 0, 1, 2, 3, 4$. I have found[1] that four ζ are insufficient to account for the three known light generations and their features. The way generations arise in this context comes by multiplying traits by neutral products of other traits. For instance a person can be pessimistic or (pessimistic $\times$ sad-happy) or (pessimistic $\times$ aggressive-submissive) and so on. With N ζ one potentially encounters 2^{N-1} generations of a particular property by forming such neutral products; this is interesting because it suggests that any anticommuting property scheme produces an even number of generations. This is not a cause for panic; all we know at present is that there exist 3 light neutrinos, so there may be other (possibly sterile) heavy neutrinos accompanied by other quarks and charged leptons.

The proposal therefore is to append a set of five complex anticommuting coordinates ζ^μ to space-time x^m, which are to be associated with property. This is in contrast to traditional brane-string schemes which append yet unseen bosonic extra degrees of freedom to space-time. Now one of the most interesting aspects of fermionic degrees of freedom is that they act oppositely to bosonic ones in several respects: we are familiar with sign changes in commutation relations, statistical formulae and, most importantly, quantum loop contributions; but less well-known is that in certain group theoretical representations[3] the $SO(2N)$ Casimirs are continuations to negative N of $Sp(2N)$ Casimirs so that anticommuting coordinates *effectively subtract* dimensions. The opposite (to bosons) loop sign is extra confirmation of this statement and this is put to great use in standard supersymmetry (SUSY) in order to resolve the fine-tuning problem. So it is not inconceivable that with a correct set of anticommuting property coordinates ζ appended to x we might end up with zero net dimensions as it was presumably before the BIG BANG; at the very least that we may arrive at a scheme where fermions cancel quantum effects from bosons as in SUSY. For this it is not necessary to embrace all the tenets of standard SUSY; indeed we shall contemplate an edifice where property coordinates are Lorentz scalar. In that respect they are similar to the variables occurring in BRST transformations for quantized vector gauge-fixed models — without implying any violation of the spin-statistics theorem for normal physical fields.

2. Property Superfields

Now let us get down to the nuts and bolts of the construction and find out what the edifice will look like. We might be tempted to match the four space-time x^m by four ζ^μ, but as we have shown elsewhere that is not enough – lepton generations do not ensue. However, all is not lost: we can add 5 ζ^μ (which are sufficient), but only take half the states to get correct statistics. Associate each Lorentz scalar anticommuting numbers with a 'property' or 'trait'; this invites us to consider symmetry groups[4] like $SU(5)$ or $Sp(10)$ or $SO(10)$ to reshuffle the properties. (That these popular groups pop up is probably no accident.) The only way I have found to obtain the well-established quantum numbers of the fundamental particle spectrum as superpositions over traits is to postulate the following charge Q and fermion number F assignments,

$$Q(\zeta^{0,1,2,3,4}) = (0, 1/3, 1/3, 1/3, -1); \quad F(\zeta^{0,1,2,3,4}) = (1, -1/3, -1/3, -1/3, 1).$$

Other properties are to built up as composites of these. For example ζ^4 may be identified as a negatively charged lepton, but then so can its product with neutral combinations like $\bar{\zeta}_0\zeta^0$, $\bar{\zeta}_i\zeta^i$,... strongly suggesting how generations can arise in this framework. We'll return to this shortly.

Since the product of an even number of ζ is a (nilpotent) commuting property a Bose superfield Φ should be a Taylor series in even powers of $\zeta, \bar{\zeta}$. Similarly a (Bose overall) superfield which encompasses fermions Ψ_α will be a series in odd powers of $\zeta, \bar{\zeta}$ — up to the 5th:

$$\Phi(x, \zeta, \bar{\zeta}) = \sum_{even\, r+\bar{r}} (\bar{\zeta})^{\bar{r}} \phi_{(\bar{r}),(r)} (\zeta)^r;$$

$$\Psi_\alpha(x, \zeta, \bar{\zeta}) = \sum_{odd\, r+\bar{r}} (\bar{\zeta})^{\bar{r}} \psi_{\alpha(\bar{r}),(r)} (\zeta)^r.$$

So far as labels μ on ζ^μ go, we can characterise

- label 0 as neutrinicity
- labels 1 - 3 as (down) chromicity
- label 4 as charged leptonicity

You will notice that these expansions produce too many states for ψ_α and ϕ, viz. 512 properties in all, so they demand cutting down. An obvious tactic is to associate conjugation (^{c}) with the operation $\zeta \leftrightarrow \bar{\zeta}$

$$\psi_{(r),(\bar{r})} = \psi^{(c)}_{(\bar{r}),(r)},$$

corresponding to reflection along the main diagonal when we expand Ψ as per the table below. Indeed if we suppose that all fermion field components are left-handed the conjugation/reflection operation then automatically includes right-handed particle states in the form of left-handed antiparticle states. Even so there remain too many components and we may wish to prune more. One strategy is to notice that under reflection about the *cross-diagonal* the F and Q quantum numbers are not altered. We shall call this cross-diagonal reflection a *duality* ($^\times$) transformation. For example,

$$(\bar{\zeta}_\alpha \zeta^\mu \zeta^\nu)^\times = \frac{1}{3!}\epsilon^{\rho\sigma\tau\mu\nu}\bar{\zeta}_\rho\bar{\zeta}_\sigma\bar{\zeta}_\tau . \frac{1}{4!}\epsilon_{\alpha\beta\gamma\delta\epsilon}\zeta^\beta\zeta^\gamma\zeta^\delta\zeta^\epsilon .$$

By imposing the antidual reflection symmetry: $\psi_{(r),(\bar{r})} = -\psi_{(5-\bar{r}),(5-r)}$ we roughly halve the remaining number of components. So whereas previously we had the separate set of neutrino states, for instance,

$$\zeta^0, \zeta^0(\bar{\zeta}_4\zeta^4), \zeta^0(\bar{\zeta}_i\zeta^i), \zeta^0(\bar{\zeta}_4\zeta^4)(\bar{\zeta}_i\zeta^i), \zeta^0(\bar{\zeta}_i\zeta^i)^2, (\bar{\zeta}_i\zeta^i)^3, \zeta^0(\bar{\zeta}_4\zeta^4)(\bar{\zeta}_i\zeta^i)^2, \zeta^0(\bar{\zeta}_4\zeta^4)(\bar{\zeta}_i\zeta^i)^3$$

antiduality sifts out half the combinations, namely:

$$\zeta^0[1 - (\bar{\zeta}_4\zeta^4)(\bar{\zeta}_i\zeta^i)^3/6], \quad \zeta^0[\bar{\zeta}_4\zeta^4) - (\bar{\zeta}_i\zeta^i)^2/2],$$

$$\zeta^0[(\bar{\zeta}_i\zeta^i) - (\bar{\zeta}_4\zeta^4)(\bar{\zeta}_j\zeta^j)^2/2], \quad \zeta^0[(\bar{\zeta}_i\zeta^i)(\bar{\zeta}_4\zeta^4) - (\bar{\zeta}_j\zeta^j)^2/2].$$

In particular as $\bar{\zeta}_0\bar{\zeta}_4\zeta^1\zeta^2\zeta^3$ and $\bar{\zeta}_4\zeta^0\zeta^1\zeta^2\zeta^3$ are self-dual, imposing antiduality eliminates these unwanted states, a good thing since they respectively have $F = 3$ & $Q = -2$. Applying antiduality and focussing on left-Ψ, the resulting square contains the following varieties of up (U), down (D), charged lepton (L) and neutrinos (N), where the subscript distinguishes between repetitions. In the $\zeta\bar{\zeta}$ expansion table, × are duals, * are conjugates:

$r\backslash\bar{r}$	0	1	2	3	4	5
0		L_1, N_1, D^c_5		L^c_5, D_1, U_1		
1	*		$L_{2,3}, N_{2,3}, D^c_{3,6,7}, U^c_3$		L^c_6, D_2, U_2	
2		*		$L_4, N_4, D^c_{4,8}, U^c_4$		×
3	*		*		×	
4		*		*		×
5	*		*		*	

Observe that colour singlet and triplet fermions come in 4s, 6s and 8s which comfortably contain the known three generations. However you will see that whereas U, D in the first and second generation are bona fide weak isospin doublets, the third and fourth family U, D are accompanied by another exotic colour triplet quark, call it X say, having charge $Q = -4/3$, and make up a weak isospin triplet. So this is a departure from the standard model! Specifically the weak isospin generators are:

$$T_+ = \zeta^0\partial_4 - \bar{\zeta}_4\bar{\partial}^0, \quad T_- = \zeta^4\partial_0 - \bar{\zeta}_0\bar{\partial}^4;$$

$$2T_3 = [T_+, T_-] = \zeta^0\partial_0 - \zeta^4\partial_4 + \bar{\zeta}_4\bar{\partial}^4 - \bar{\zeta}_0\bar{\partial}^0, \text{ so we meet}$$

$$\text{doublets like } (N_1, L_1), \quad (U_1, D_1) \sim (\zeta^0, \zeta^4), \quad (-\bar{\zeta}_4, \bar{\zeta}_0)$$

$$\text{singlets like } L_5\ D_5 \sim (\bar{\zeta}_0\zeta^0 + \bar{\zeta}_4\zeta^4), \quad (\bar{\zeta}_0\zeta^0\bar{\zeta}_4\zeta^4),$$

$$\text{triplets like } (U_3, D_3, X_3) \sim (-\bar{\zeta}_4\zeta^0, [\bar{\zeta}_0\zeta^0 - \bar{\zeta}_4\zeta^4]/\sqrt{2}, \bar{\zeta}_0\zeta^4).$$

3. Exotic Particles, Generations and the Mass Matrix

The U-states arise from combinations like $\zeta^i\zeta^j\zeta^0$ lying in the $\overline{10}$-fold SU(5) combination $\zeta^\lambda\zeta^\mu\zeta^\nu$ and as $\bar{\zeta}_k\bar{\zeta}_4\zeta^0$. (Note that $\zeta^0\zeta^4\zeta^k$ has the exotic value $F = 5/3$ and cannot be identified with U^c.) The D-states occur similarly, as do the N's and L's. However, observe that $L_{5,6} \sim \bar{\zeta}_3\bar{\zeta}_2\bar{\zeta}_1$ and $D_{5,6,7,8} \sim \bar{\zeta}_k$ are nominally weak isosinglets, which again *differs* from the standard model. These mysterious states are definitely charged but do not possess any weak interactions, so their behaviour is very curious. One might even regard $L_{5,6}$ like colourless charged baryons (antiproton-like) but really until one sees how these states mix with the usual weak isodoublet leptons it is dangerous to label them one thing rather than another without further research.

'Pentaquarks' such as $\Theta^+ \sim uudd\bar{s}$ & $\Xi^{--} \sim ddss\bar{u}$ were recently discovered(?) with quite narrow widths and many people have advanced models to describe these new resonances as well as tetraquark mesons. But who is to say unequivocally that they are not composites of ordinary quarks and another U-quark or other D-family quarks which my scheme indicates? Further, I get a fourth neutrino, which could be essentially sterile and might help explain the mystery of ν masses. This is clearly fertile ground for investigation and I have only scratched the surface here. Possibly conflicts with experiment may arise that will ultimately invalidate the entire property values scheme.

One of the first matters to be cleared up is the mass matrix and flavour mixing which affects quarks as well as leptons. If we assume it is due to a Higgs Φ field's expectation values, there are nine colourless possibilities having $F = Q = 0$ lying in an antidual boson superfield:

- one $\phi_{(0)(0)} = \langle\phi\rangle = M$
- one $\phi_{(0)(4)} = \langle\phi_{1234}\rangle = H$ complex
- three $\phi_{(1)(1)} = \langle\phi^0_0, \phi^4_4, \phi^i_i\rangle = A, B, C$
- four $\phi_{(2)(2)} = \langle\phi^{04}_{04}, \phi^{0k}_{0k}, \phi^{4k}_{4k}, \phi^{ij}_{ij}\rangle = D, E, F, G,$

others being related by duality. With nine $\langle\phi\rangle$ the mass matrix calculation already becomes a difficult task! To ascertain what happens, consider the subset of quarks involving U_{1-4} and $D_{1-4,7}$ interacting with anti-selfdual superHiggs.

The Lagrangian $\int d^5\zeta d^5\bar{\zeta}\ \bar{\Psi}\langle\Phi\rangle\Psi$ produces U and emasculated D mixing matrices:

$$2M(U) \rightarrow \begin{pmatrix} 2M+F/\sqrt{3} & B+C/\sqrt{3} & -H^* & 0 \\ B+C/\sqrt{3} & 2M & 0 & 0 \\ -H & 0 & 2M+G/\sqrt{3} & 2C/\sqrt{3} \\ 0 & 0 & 2C/\sqrt{3} & 2M \end{pmatrix}$$

$$2M(D) \rightarrow \begin{pmatrix} 2M+E/\sqrt{3} & A+C/\sqrt{3} & -H^* & 0 & 0 \\ A+C/\sqrt{3} & 2M & 0 & 0 & 0 \\ -H & 0 & 2M+G/\sqrt{3} & 2C/\sqrt{3} & (E-F)/\sqrt{3} \\ 0 & 0 & 2C/\sqrt{3} & 2M & A-B \\ 0 & 0 & -(E+F)/\sqrt{3} & A-B & 2M-D \end{pmatrix}.$$

Similar expressions can be found for the leptons and quarks. Such matrices have to be diagonalised via unitary transformations $V(U)$ & $V(D)$. However the coupling of the weak bosons is *different* for isodoublets $U_{1,2}$ $D_{1,2}$ and isotriplets $U_{3,4}$ $D_{3,4}$ (and zero for the weak isosinglet D_7). So we can't just evaluate $V^{-1}(U)V(D)$ for the unitary CKM matrix now. Rather the weak interactions connecting U & D mass-diagonalised quarks will not be quite unitary (because of the different coupling factors):

$$L_{\rm weak}/g_w = W^+(\bar{U}_1D_1+\bar{U}_2D_2)+\sqrt{2}W^+(\bar{U}_3D_3+\bar{U}_4D_4+\bar{D}_3X_3+\bar{D}_4X_4)$$

$$+W^-(\bar{D}_1U_1+\bar{D}_2U_2)+\sqrt{2}W^-(\bar{D}_3U_3+\bar{D}_4U_4+\bar{X}_3D_3+\bar{X}_4D_4)+W^3 \text{ terms.}$$

So this is another departure from the standard picture: nonunitarity of the 3×3 CKM matrix is a test of the scheme. Also CP violation is an intrinsic feature of the property formalism because H is naturally complex, unlike the other expectation values $A, B...M$. A more realistic attempt for getting the quark and lepton masses would be to abandon antiduality in the Higgs sector and use all 18 expectation values; otherwise it may prove impossible to cover the 12 or more orders of magnitude all the way from the electron neutrino to the top quark (and higher).

4. Generalized Relativity

We know that gauge fields can transport/communicate property from one place to another so where are they? Maybe one can mimic the SUSY procedure and supergauge the massless free action for Ψ, without added complication of spin. But there is a more compelling way, which has the benefit of incorporating gravity. In terms of an extended coordinate $X^M=(x^m,\zeta^\mu)$, construct a fermionic version of Kaluza-Klein (KK) theory,[5] this time without worrying about infinite modes which arise from squeezing normal bosonic coordinates. These are the significant points of such an approach:

- One must introduce a fundamental length Λ in the extended X, as property ζ has no dimensions; maybe this is the gravity scale $\kappa = \sqrt{8\pi G_N}$?
- Gravity (plus gauge field products) fall within the $x - x$ sector, gauge fields in $x - \zeta$ and the Higgs scalars must form a matrix in $\zeta - \zeta$,
- Gauge invariance is connected with the number of ζ so $SU(5)$ or $Sp(10)$ *or perhaps a subgroup* are indicated,
- There is no place for a gravitino as spin is absent (ζ are Lorentz scalar),
- There are necessarily a small finite number of modes,
- Weak left-handed SU(2) is associated with rotations of ζ **not** $\bar{\zeta}$ so may have something to with ζ-analyticity.

The real metric specifies the separation in location as well as property: it tells us how 'far apart' and 'different in type' two events are. Setting $\bar{\zeta}^{\bar{\mu}} \equiv \bar{\zeta}_\mu$,

$$ds^2 = dx^m dx^n G_{nm} + dx^m d\zeta^\nu G_{\nu m} + dx^n d\bar{\zeta}^{\bar{\mu}} G_{\bar{\mu}n} + d\bar{\zeta}^{\bar{\mu}} d\zeta^\nu G_{\bar{\mu}\nu}$$

where the tangent space limit corresponds to Minkowskian

$$G_{ab} \to I_{ab} = \eta_{ab},\ G_{\bar{\alpha}\beta} \to I_{\bar{\alpha}\beta} = \Lambda^2 \delta_{\alpha\beta},$$

multiplied at least by $(\bar{\zeta}\zeta)^5$ — to arrange correct property integration. Proceeding to curved space the components should contain the force fields, leading one to a 'superbein'

$$\bar{E}_M^A = \begin{pmatrix} e_m{}^a & i\Lambda(A_m)_\mu^\alpha \zeta^\mu \\ 0 & \Lambda\delta_\mu^\alpha \end{pmatrix},$$

and the metric "tensor" arising from

$$ds^2 = dx^m dx^n g_{nm} + 2\Lambda^2[d\bar{\zeta}^{\bar{\mu}} - idx^m \bar{\zeta}^{\bar{\kappa}} (A_m)_{\bar{\kappa}}^{\bar{\mu}}]\delta_{\bar{\mu}\nu}[d\zeta^\nu + idx^n (A_n)_\lambda^\nu \zeta^\lambda];$$

$$g_{mn} = e_m^a e_n^b \eta_{ab}.$$

Gauge symmetry corresponds to the special change $\zeta^\mu \to \zeta'^\mu = [\exp(i\Theta(x))]_\nu^\mu \zeta^\nu$ with $x' = x$. Given the standard transformation law

$$G_{\zeta m}(X) = \frac{\partial X'^R}{\partial x^m}\frac{\partial X'^S}{\partial \zeta} G'_{SR}(X')(-1)^{[R]} = \frac{\partial \zeta'}{\partial \zeta} G'_{\zeta m} - \frac{\partial \bar{\zeta}'}{\partial x^m}\frac{\partial \zeta'}{\partial \zeta} G'_{\zeta\bar{\zeta}},$$

this translates into the usual gauge variation (a matrix in property space),

$$A_m(x) = \exp(-i\Theta(x))[A'_m(x) - i\partial_m]\exp(i\Theta(x)).$$

The result is consistent with other components of the metric tensor but does not fix what (sub)group is to be gauged in property space although one most certainly expects to take in the nonabelian colour group and the abelian electromagnetic group, so as to agree with physics.

5. Some Notational Niceties

When dealing with commuting and anticommuting numbers within a single coordinate framework $X^M = (x^m, \zeta^\mu)$ one has to be **exceedingly** careful with the order of quantities and of labels. I cannot stress this enough. (It took me six months and much heartache to get the formulae below correct.) For derivatives the rule is $dF(X) = dX^M(\partial F/\partial X^M) \equiv dX^M \partial_M F$, **not** with the dX on the right, and for products of functions: $d(FG..) = dF\,G + F\,dG + ...$ Coordinate transformations read $dX'^M = dX^N(\partial X'^M/\partial X^N)$, and *in that particular order.* Since $ds^2 = dX^N dX^M G_{MN}$, the symmetry property of the metric is $G_{MN} = (-1)^{[M][N]}G_{NM}$. Let $G^{LM}G_{MN} = \delta^L_N$ for the inverse metric, so $G^{MN} = (-1)^{[M]+[N]+[M][N]}\,G^{NM}$. The notation here is $[M] = 0$ for bosons and 1 for fermions.

Changing coordinate system from X to X', we have to be punctilious with signs and orders of products, things we normally never fuss about; the correct transformation law is

$$G_{NM}(X) = \left(\frac{\partial X'^R}{\partial X^M}\right)\left(\frac{\partial X^S}{\partial X^N}\right) G'_{SR}(X')\,(-1)^{[N]([R]+[M])}.$$

Transformation laws for contravariant and covariant vectors read:

$$V'^M(X') = V^R(X)\left(\frac{\partial X'^M}{\partial X^R}\right) \quad \text{and} \quad A'_M(X') = \left(\frac{\partial X^R}{\partial X'^M}\right) A_R(X),$$

in the order stated. Thus the invariant contraction is

$$V'^M(X')A'_M(X') = V^R(X)A_R(X) = (-1)^{[R]}A_R(X)V^R(X).$$

The inverse metric G^{MN} can be used to raise and lower indices as well as forming invariants, so for instance $V_R \equiv V^S G_{SR}$ and $V'^R V'^S G'_{SR} = V^M V^N G_{NM}$.

The next issue is covariant differentiation; we insist that $A_{M;N}$ should transform like T_{MN}, viz.

$$T'_{MN}(X') = (-1)^{[S]+[N])[R]}\left(\frac{\partial X^R}{\partial X'^M}\right)\left(\frac{\partial X^S}{\partial X'^N}\right) T_{RS}(X).$$

After some work we find that

$$A_{M;N} = (-1)^{[M][N]}A_{M,N} - A^L\Gamma_{\{MN,L\}},$$

where the connection is given by

$$\begin{aligned}\Gamma_{\{MN,L\}} &\equiv [(-1)^{([L]+[M])[N]}G_{LM,N} + (-1)^{[M][L]}G_{LN,M} - G_{MN,L}]/2\\ &= (-1)^{[M][N]}\Gamma_{\{NM,L\}}.\end{aligned}$$

Another useful formula is the raised connection

$$\Gamma_{MN}{}^K \equiv (-1)^{[L]([M]+[N])}\Gamma_{\{MN,L\}}G^{LK} = (-1)^{[M][N]}\Gamma_{NM}{}^K,$$

whereupon one may write

$$A_{M;N} = (-1)^{[M][N]}A_{M,N} - \Gamma_{MN}{}^L A_L.$$

Similarly one can show that for double index tensors the true differentiation rule is

$$T_{LM;N} \equiv (-1)^{[N]([L]+[M])} T_{LM,N} - (-1)^{[M][N]} \Gamma_{LN}{}^{K} T_{KM} - (-1)^{[L]([M]+[N]+[K])} \Gamma_{MN}{}^{K} T_{LK}.$$

As a nice check, the covariant derivative of the metric properly vanishes:

$$G_{LM;N} \equiv (-1)^{[N]([L]+[M])} G_{LM,N} - (-1)^{[L][M]} \Gamma_{\{LN,M\}} - \Gamma_{\{MN,L\}} \equiv 0.$$

Moving on to the Riemann curvature we form doubly covariant derivatives:

$$A_{K;L;M} - (-1)^{[L][M]} A_{K;M;L} \equiv (-1)^{[K]([L]+[M])} R^{J}{}_{KLM} A_{J}$$

where one discovers that

$$\begin{aligned} R^{J}{}_{KLM} \equiv\ & (-1)^{[K][M]} (\Gamma_{KM}{}^{J})_{,L} - (-1)^{[L]([K]+[M])} (\Gamma_{KL}{}^{J})_{,M} \\ & + (-1)^{[M]([K]+[L])+[K][L]} \Gamma_{KM}{}^{N} \Gamma_{NL}{}^{J} - (-1)^{[K]([M]+[L])} \Gamma_{KL}{}^{N} \Gamma_{NM}{}^{J}. \end{aligned}$$

Evidently, $R^{J}{}_{KLM} = -(-1)^{[L][M]} R^{J}{}_{KML}$ and, less obviously, the cyclical relation takes the form

$$(-1)^{[K][L]} R^{J}{}_{KLM} + (-1)^{[L][M]} R^{J}{}_{LMK} + (-1)^{[M][K]} R^{J}{}_{MKL} = 0.$$

The fully covariant Riemann tensor is $R_{JKLM} \equiv (-1)^{([J]+[K])[L]} R^{N}{}_{KLM} G_{NJ}$ with pleasing features:

$$\begin{aligned} R_{JKLM} &= -(-1)^{[L][M]} R_{JKML} = -(-1)^{[J][K]} R_{KJLM}, \\ 0 &= (-1)^{[J][L]} R_{JKLM} + (-1)^{[J][M]} R_{JLMK} + (-1)^{[J][K]} R_{JMKL} \\ R_{JKLM} &= (-1)^{([J]+[K])([L]+[M])} R_{LMJK}. \end{aligned}$$

Finally proceed to the Ricci tensor and scalar curvature:

$$\begin{aligned} R_{KM} &\equiv (-1)^{[J]+[K][L]+[J]([K]+[M])} G^{LJ} R_{JKLM} \\ &= (-1)^{[L]([K]+[L]+[M])} R^{L}{}_{KLM} = (-1)^{[K][M]} R_{MK}, \end{aligned}$$

$$R \equiv G^{MK} R_{KM}.$$

6. Curvatures of Space-Time-Property

When Einstein produced his general theory of relativity with Grossmann he wrote all his expressions in terms of real variables. In order to avoid any confusion with complex variables, we shall copy him by writing everything in terms of real coordinates ξ, η rather than complex $\zeta = (\xi + i\eta)/\sqrt{2}$. (Note that the real invariant is $\bar{\zeta}\zeta = i\xi\eta$ and that a phase transformation of ζ corresponds to a real rotation in (ξ, η) space.) For simplicity consider just one extra pair, rather than five pairs, and the following two examples, which are complicated enough as it is.

(1) Decoupled property and space-time, but both curved:

$$\begin{aligned} ds^2 &= dx^m dx^n G_{nm}(x,\xi,\eta) + 2id\xi d\eta G_{\eta\xi}(x,\xi,\eta) \\ &\equiv dx^m dx^n g_{nm}(x)(1+if\xi\eta) + 2i\Lambda^2 d\xi d\eta(1+ig\xi\eta) \end{aligned}$$

from which we can read off the metric components ($G^{LM}G_{MN} \equiv \delta^L_N$)

$$\begin{pmatrix} G_{mn} & G_{m\xi} & G_{m\eta} \\ G_{\xi n} & G_{\xi\xi} & G_{\xi\eta} \\ G_{\eta n} & G_{\eta\xi} & G_{\eta\eta} \end{pmatrix} = \begin{pmatrix} g_{mn}(1+if\xi\eta) & 0 & 0 \\ 0 & 0 & -i\Lambda^2(1+ig\xi\eta) \\ 0 & i\Lambda^2(1+ig\xi\eta) & 0 \end{pmatrix},$$

$$\begin{pmatrix} G^{lm} & G^{l\xi} & G^{l\eta} \\ G^{\xi m} & G^{\xi\xi} & G^{\xi\eta} \\ G^{\eta m} & G^{\eta\xi} & G^{\eta\eta} \end{pmatrix} = \begin{pmatrix} g^{lm}(1-if\xi\eta) & 0 & 0 \\ 0 & 0 & -i(1-ig\xi\eta)/\Lambda^2 \\ 0 & i(1-ig\xi\eta)/\Lambda^2 & 0 \end{pmatrix}.$$

The non-zero connections in the property sector are

$$\Gamma^{\xi}_{\xi\eta} = -\Gamma_{\eta\xi}{}^{\xi} = ig\xi, \quad \Gamma_{\xi\eta}{}^{\eta} = -\Gamma_{\eta\xi}{}^{\eta} = ig\eta$$

so

$$\begin{aligned} R^{\eta}{}_{\xi\eta\eta} &= -2ig(1+ig\xi\eta) = -R^{\xi}{}_{\eta\xi\xi} \\ R^{\xi}{}_{\xi\xi\eta} &= -ig(1+ig\xi\eta) = -R^{\eta}{}_{\eta\eta\xi} \\ \text{so } R_{\xi\eta} &= -R_{\eta\xi} = 3ig(1+ig\xi\eta). \end{aligned}$$

Consequently the total curvature is given by

$$R = G^{mn}R_{nm} + 2G^{\eta\xi}R_{\xi\eta} = R^{(g)}(1-if\xi\eta) - 6g/\Lambda^2. \tag{1}$$

Since $\sqrt{-G..} = -i\Lambda^2\sqrt{-g..}(1+2if\xi\eta)(1+ig\xi\eta)$, we obtain an action

$$I \equiv \frac{1}{2\Lambda^4}\int R\sqrt{G..}\, d^4x d\eta d\xi = \frac{1}{2\kappa^2}\int d^4x\sqrt{-g..}\left[R^{(g)} + \lambda\right]$$

where $\kappa^2 \equiv 8\pi G_N = \Lambda^2/(f+g)$, $R^{(g)}$ is the standard gravitational curvature and $\lambda = -6g(2f+g)/\Lambda^2(f+g)$ corresponds to a cosmological term.

(2) Our second example leaves property space flat (in the η, ξ sector) but introduces a $U(1)$ gauge field A, governed by the metric,

$$\begin{pmatrix} G_{mn} & G_{m\xi} & G_{m\eta} \\ G_{\xi n} & G_{\xi\xi} & G_{\xi\eta} \\ G_{\eta n} & G_{\eta\xi} & G_{\eta\eta} \end{pmatrix} = \begin{pmatrix} g_{mn}(1+if\xi\eta)+2i\Lambda^2\xi A_m A_n\eta & i\Lambda^2 A_m\xi & i\Lambda^2 A_m\eta \\ i\Lambda^2 A_n\xi & 0 & -i\Lambda^2 \\ i\Lambda^2 A_n\eta & i\Lambda^2 & 0 \end{pmatrix}.$$

Simplify the analysis somewhat by going to flat (Minkowski) space first as there are then fewer connections. After some work ($F_{mn} \equiv A_{m,n} - A_{n,m}$) one obtains,

$$\begin{aligned} &\Gamma_{\xi\eta}{}^{\xi} = \Gamma_{\xi\eta}{}^{\eta} = \Gamma_{\xi\eta}{}^{k} = 0, \\ &\Gamma_{m\xi}{}^{\xi} = \Gamma_{m\eta}{}^{\eta} = i\Lambda^2 A^l F_{lm}\xi\eta/2, \quad \Gamma_{m\xi}{}^{\eta} = -\Gamma_{m\eta}{}^{\xi} = A_m, \\ &\Gamma_{m\xi}{}^{l} = i\Lambda^2 F^l{}_m\xi/2, \quad \Gamma_{m\eta}{}^{l} = i\Lambda^2 F^l{}_m\eta/2, \\ &\Gamma_{mn}{}^{\xi} = -A_m A_n\xi - (A_{m,n} + A_{n,m})\eta/2, \\ &\Gamma_{mn}{}^{\eta} = -A_m A_n\eta + (A_{m,n} + A_{n,m})\xi/2, \\ &\Gamma_{mn}{}^{k} = i\Lambda^2(A_m F^k{}_n + A_n F^k{}_m)\xi\eta. \end{aligned}$$

Other Christoffel symbols can be deduced through symmetry of indices. Hence

$$R_{km} = R^l{}_{klm} - R^\xi{}_{k\xi m} - R^\eta{}_{k\eta m} = -i\Lambda^2(A_{k,l} + A_{l,k})F^l{}_m\xi\eta/2 + \text{total der.}$$

$$R_{k\xi} = R^l{}_{kl\xi} + R^\xi{}_{k\xi\xi} + R^\eta{}_{k\eta\xi} = i\Lambda^2[F^l{}_{k,l}\xi/2 + A^l F_{k,l}\eta] + \text{total der.}$$

$$R_{k\eta} = R^l{}_{kl\eta} + R^\xi{}_{k\xi\eta} + R^\eta{}_{k\eta\eta} = i\Lambda^2[F^l{}_{k,l}\eta/2 - A^l F_{k,l}\xi] + \text{total der.}$$

$$R_{\xi\eta} = -\Lambda^4 F_{kl}F^{lk}\xi\eta.$$

Then covariantize by including the gravitational component $g_{mn}(1 + if\xi\eta)$ to end up with the total curvature:

$$\begin{aligned} R &= G^{mn}R_{nm} + 2G^{m\xi}R_{\xi m} + 2G^{m\eta}R_{\eta m} + 2G^{\eta\xi}R_{\xi\eta} \\ &\to R^{(g)} - 3i\Lambda^2 g^{km}g^{ln}F_{kl}F_{nm}\xi\eta/2. \end{aligned}$$

Finally rescale A to identify the answer as electromagnetism + gravitation:

$$\int R\sqrt{-G..}d^4x d\eta d\xi/4\Lambda^4 = \int d^4x\sqrt{-g..}\left[R^{(g)}/2\kappa^2 - F^{kl}F_{kl}/4\right],$$

where $\kappa^2 = \Lambda^2/f = 8\pi G_N$. It is a nice feature of the formalism that the gauge field Lagrangian arises from space-property terms — like the standard K-K model (from the tie-up between ordinary space-time and the fifth dimension).

Where do we go from these two examples? Well, some generalizations come to mind:

- replace property couplings f and g by two fields (dilaton and Higgs),
- combine the two models; this should lead to gravity + em + cosmic const.,
- extend fully to five ζ; it is easy enough to incorporate the standard gauge model and one can even entertain a GUT $SU(5)$ of some ilk,
- work out the particle mass spectrum from all the $\langle\phi\rangle$.

In conclusion I have tried to persuade you that connecting properties with anticommuting coordinates is an idea worth exploring. The concept is intriguing, more tangible and less outlandish than other schemes based on extended space-time structures. In the end the idea may turn out to be completely wrong but for now it has a lot going for it. Eventually experiment will decide.

Acknowledgments

A considerable amount of the material has been drawn from previous papers written with several collaborators including Peter Jarvis, Ruibin Zhang, Roland Warner and Martin White. I owe them my thanks for all their valuable insights and help.

References

1. R. Delbourgo, *J. Phys. A* **39**, 5175 (2006).
2. R. Delbourgo, *J. Phys. A* **39**, 14735 (2006).
3. A. McKane, Phys. Lett. *A76*, 22 (1980);
   P. Cvitanovic and A.D. Kennedy, *Phys. Scripta* **26**, 5 (1982);
   I.G. Halliday and R.M. Ricotta, *Phys. Lett.* **B193**, 241 (1987);
   G.V. Dunne, *J. Phys.* **A22**, 1719 (1989).
4. P.D. Jarvis and M. White, *Phys. Rev.* **D43**, 4121 (1991);
   R. Delbourgo, P.D. Jarvis, R.C. Warner, *Aust. J. Phys.* **44**, 135 (1991).
5. R. Delbourgo, S. Twisk and R. Zhang, *Mod. Phys. Lett.* **A3**, 1073 (1988);
   P. Ellicott and D.J. Toms, *Class. Quant. Grav.* **6**, 1033 (1989);
   R. Delbourgo and M. White, *Mod. Phys. Lett.* **A5**, 355 (1990).

INTERACTING BOSE GAS CONFINED BY AN EXTERNAL POTENTIAL

G. GNANAPRAGASAM and M. P. DAS*

Department of Theoretical Physics, The Australian National University, Canberra, ACT 0200, Australia
**E-mail: mpd105@rsphysse.anu.edu.au*

Interacting Bose gas confined by an external potential is studied using Green functions in spectral representation. The calculation is presented transparently using the equation of motion method. With this, the interplay between the condensed and the non-condensed atoms is inevitably seen. An expression for the condensate number at finite temperature is obtained in the lowest and first orders, from which depletion of bosons from the ground state is qualitatively analyzed. Finally, we discuss the behaviour of the specific heat of a trapped interacting Bose gas in the quasi-continuum limit.

Keywords: Bose Condensation; Interaction; External Confinement; Green Functions; Beyond Mean Field; Condensate Number.

1. Introduction

The successful observation of Bose-Einstein Condensation(BEC) in dilute gases[1,2] of magnetically trapped atomic vapours, has given rise to a deluge of both experimental and theoretical studies on confined systems. A great deal of the current understanding of the phenomenon of BEC can be found from references [3–7]. The ground state properties of trapped interacting Bose gas at zero temperature, has been so far successfully explained by the Gross-Pitaevskii(GP)[8,9] equation. Many properties of the condensate such as its size, shape and energy can be obtained from the stationary solutions of the GP equation.[10,11] Due to the presence of two-body interactions, not all the particles are in the zero-momentum state. This effect is known as "quantum depletion". But for systems with low-density, this so called "depletion" was neglected as in the work of Bogoliubov[12] and in the pseudopotential method of Lee, Huang and Yang.[13] The effect of depletion was taken into account by Beliaev[14] and was extended by Hugenholtz and Pines[15] using field-theoretic Green function methods. In most experiments that have been carried out, the condensate depletion is of the order of 1%. However, one of the experiments on ^{85}Rb near a Feshbach resonance, achieved[16] large values of scattering length corresponding to a depletion of 10%, thus opening up new channels to go beyond the existing mean-field theories.

---

*Speaker.

At finite temperatures and in the presence of strong interactions, it becomes necessary to take into account, the effect of the non-condensate on condensed atoms. The major difference between a Bose-condensed gas at low T and a Bose-condensed liquid like superfluid ^4He is that in the latter, the condensate fraction is about 10%[17](even at extremely low temperature and pressure), and thus it is crucial to include the effect of the non-condensate atoms, which are neglected in the simple Bogoliubov approximation. Finite-temperature calculations for an interacting Bose gas both in the homogeneous limit[18–20] and in trapped gases[21–24] have been studied earlier. Such calculations are much more complicated than the $T = 0$ case due to correlations induced by the Bose broken symmetry and also due to the presence of finite number of thermally excited (depleted) atoms.

In this paper, we give a general theory of a confined interacting Bose system at finite temperatures. We study the static properties of the condensate under external confinement, in the presence of both repulsive and attractive interatomic interactions. This is done using the double time temperature Green functions which were first introduced by Bogoliubov and Tyablikov. Green functions in statistical physics have been found to be extremely useful in the last few decades.[25–28] This has made possible the developement of methods that could not be explained by the ordinary perturbation theory. In such many-body problems, the single particle Green function[14,29,30] determines the essential features of the system like the quasi-particle energy spectrum, ground state occupation etc.

The condensate acts as a reservoir for atoms to move in and out of it and as a result the non-condensate atoms involve in scattering processes, with the atoms in the condensate. From this, we see that the condensate number is not conserved and as an immediate consequence we introduce the anomalous Green function which is a characteristic feature of the system that has undergone a phase transition to the condensed state.[31] The anomalous Green function represents either two particles going into or out of the condensate, that is, either to make a pair creation or pair annihilation respectively. In such an intuitive formalism, we take into account both the non-zero density of non-condensate atoms and the anomalous density. Thus, we explicitly see the inevitable interplay between the condensate and the non-condensate components, which makes the theory applicable even to strongly interacting systems and hence we go beyond the usual mean-field theories.

2. Hamiltonian

We now consider a 3D system of N interacting bosons trapped by an external potential V_{trap}. The total Hamiltonian for the system under consideration, in the second quantized form is given by:

$$H = \sum_{ij} H_{ij}^{(0)} a_i^{\dagger} a_j + \frac{1}{2} \sum_{ijkl} < ij|V^I|kl > a_i^{\dagger} a_j^{\dagger} a_k a_l \tag{1}$$

where $H_{ij}^{(0)}$, is the one-body Hamiltonian, $a^\dagger, a$ are the creation and annihilation operators which are defined through the relations:

$$a_i^\dagger\ |n_0, n_1, ..., n_i, ... > = \sqrt{n_i + 1}\ |n_0, n_1, ..., n_i + 1, ... >$$
$$a_i\ |n_0, n_1, ..., n_i, ... > = \sqrt{n_i}\ |n_0, n_1, ..., n_i - 1, ... >$$

and V^I is the two-body interaction potential.

$$H_{ij}^{(0)} = \int d\mathbf{r} \left[\phi_i^*(\mathbf{r}) \left(-\frac{\hbar^2}{2m} \nabla^2 + V_{trap}(\mathbf{r}) \right) \phi_j(\mathbf{r}) \right] \tag{2}$$

and

$$< ij|V^I|kl > = \int d\mathbf{r} d\mathbf{r}' \phi_i^*(\mathbf{r}) \phi_j^*(\mathbf{r}') V^I \phi_k(\mathbf{r}') \phi_l(\mathbf{r}) \tag{3}$$

where the ϕ's are the eigenstates of the unperturbed Hamiltonian that form a complete set of states, in general, they do not correspond to the states of the Hamiltonian given in Eq. (1).

In the ground state of the total Hamiltonian, not only the occupation number operators $N_i = a_i^\dagger a_i$, but also operators like $a_i^\dagger a_j^\dagger$ and $a_i a_j$, (the "pair Hamiltonian", the operator that creates or annihilates pairs) have non-vanishing expectation values. As a result, the total Hamiltonian contains basically three processes, namely the forward, exchange and pair scattering processes, with the remaining terms in the interaction Hamiltonian being neglected. In the earlier works,[12,32] the total Hamiltonian was diagonalized using a canonical transformation to obtain the thermodynamic quantities. Whereas here, the total Hamiltonian is separated out for the purely condensed state, the non-condensed state and the interaction between the condensate and the non-condensate. As we intend to find the ground state energy, we have not used the Bogoliubov approximation, which removes the ground state creation and annihilation operators from the Hamiltonian. Moreover, for a strongly interacting system like ^4He, in which the condensate fraction is only 10%, even at $T = 0$, Bogoliubov approximation cannot be used. Thus, taking into account all possible matrix elements, the model Hamiltonian H is :

$$\begin{aligned} H = {} & \epsilon_0 a_0^\dagger a_0 + \frac{1}{2} V_0 a_0^\dagger a_0^\dagger a_0 a_0 + \sum_i{}' \epsilon_{i0} a_i^\dagger a_0 + \sum_i{}' \epsilon_{0i} a_0^\dagger a_i + \sum_{ij}{}' \epsilon_{ij} (a_i^\dagger a_j + a_j^\dagger a_i) \\ & + \sum_i{}' V_{i000} a_i^\dagger a_0^\dagger a_0 a_0 + \sum_i{}' V_{00i0} a_0^\dagger a_0^\dagger a_i a_0 + \frac{1}{2} \sum_{ij}{}' V_{ij00} a_i^\dagger a_j^\dagger a_0 a_0 + \frac{1}{2} \sum_{ij}{}' V_{00ij} a_0^\dagger a_0^\dagger a_i a_j \\ & + 2 \sum_{ij}{}' V_{i0j0} a_i^\dagger a_0^\dagger a_j a_0 + \sum_{ijk}{}' V_{ijk0} a_i^\dagger a_j^\dagger a_k a_0 + \sum_{ijk}{}' V_{k0ij} a_k^\dagger a_0^\dagger a_i a_j + \frac{1}{2} \sum_{ijkl}{}' V_{ijkl} a_i^\dagger a_j^\dagger a_k a_l \end{aligned} \tag{4}$$

where we have used the short-hand notation: $\epsilon_0 = H_{00}^{(0)}$, $\epsilon_{0j} = H_{0j}^{(0)}$, $V_0 = < 00|V^I|00 >$, $V_{i000} = < i0|V^I|00 >$ etc. Thus the model Hamiltonian is split into fourteen terms giving the energy of the condensate, energy of the non-condensate and the correlation energy. The interaction Hamiltonian is split into nine terms representing various possible interaction processes.

3. Double-Time Temperature-Dependent Green Functions

We now, define the double-time temperature Green functions, with which we obtain an expression for the condensate number. This is done by solving the equations for the Green functions in spectral representation, using physically meaningful approximations. The equations for the Green functions are transparent and are obtained with much less labour.

Consider the thermodynamic quantity A. Its ensemble average is given by:

$$< A >= \frac{Tr A_{op} \exp(-\beta H_{op})}{Tr \exp(-\beta H_{op})} \tag{5}$$

where A_{op} is a time dependent operator which in the Heisenberg representation is given by: $A_{op}(t) = \exp(iH_{op}t)A_{op}\exp(-iH_{op}t)$ and $\beta = 1/kT$, where T is the equilibrium temperature.

The double-time temperature Green functions are defined[33,34] as:

$$\begin{aligned} G_c(t,t') = -i\theta(t-t') < A(t)B(t') > -i\theta(t'-t) < B(t')A(t) > \\ \equiv \ll A(t); B(t') \gg_c \end{aligned} \tag{6}$$

$$G_r(t,t') = -i\theta(t-t') < [A(t), B(t')] > \equiv \ll A(t); B(t') \gg_r \tag{7}$$

$$G_a(t,t') = i\theta(t'-t) < [A(t), B(t')] > \equiv \ll A(t); B(t') \gg_a \tag{8}$$

where G_c, G_r and G_a are the causal, retarded and advanced Green functions respectively as denoted by the suffices c, r and a. A and B are the Bose operators. θ is the heaviside step function. For applications in many-body physics, one normally finds it very convenient to use the retarded and advanced Green functions, as they can be analytically continued in the complex energy plane.

4. Equation of Motion

We now define the normal and the anomalous Green functions, which denote the normal and the anomalous ordering respectively, between the states m and n: $G_{mn} =\ll a_m; a_n^\dagger \gg$ and $F_{mn} =\ll a_m^\dagger; a_n^\dagger \gg$, where a_m and $a_n^\dagger$ are operators in the Heisenberg representation.

The equation of motion for G_{mn} is given by:

$$i\frac{d}{dt}G_{mn} = \delta(t-t') < [a_m; a_n^\dagger] > + \ll i\frac{da_m(t)}{dt}; a_n^\dagger \gg \tag{9}$$

In the Heisenberg representation, a_m satisfies the equation

$$i\frac{da_m(t)}{dt} = a_m H - H a_m \tag{10}$$

Thus we get:

$$i\frac{d}{dt}G_{mn} = \delta(t-t') < [a_m; a_n^\dagger] > + \ll [a_m, H]; a_n^\dagger \gg \tag{11}$$

The right hand side consists of an inhomogeneous term and a term involving another usually more complicated, double-time Green function. The equation of motion for this new Green function can then be written and the continuation of this process leads[35] to a chain of equations, which are never ending if the Hamiltonian contains interaction terms. As the chain of equations for the Green functions does not terminate, it cannot be solved exactly. The usual procedure is to decouple the hierarchy of equations by introducing physically motivated approximations that relate a higher order Green function to a lower order one.

Similarly, the equation of motion for F_{mn} is:

$$i\frac{d}{dt}F_{mn} = \delta(t-t') < [a_m^\dagger; a_n^\dagger] > + \ll [a_m^\dagger, H]; a_n^\dagger \gg . \tag{12}$$

Evaluating the commutators and after taking the time Fourier transform we get the equations of motion for G_{mn} and F_{mn} as functions of energy. From which we get the condensed state Green functions, which are

$$\begin{aligned}
EG_{00} = {} & \frac{1}{2\pi} + \epsilon_0 \ll a_0; a_0^\dagger \gg + V_0 \ll a_0^\dagger a_0 a_0; a_0^\dagger \gg + \sum_i{}' \epsilon_{0i} \ll a_i; a_0^\dagger \gg \\
& + \sum_i{}' V_{i000} \ll a_i^\dagger a_0 a_0; a_0^\dagger \gg + 2\sum_i{}' V_{00i0} \ll a_0^\dagger a_i a_0; a_0^\dagger \gg \\
& + \sum_i{}' V_{00ii} \ll a_0^\dagger a_i a_i; a_0^\dagger \gg + \sum_{i,j\neq i}{}' V_{00ij} \ll a_0^\dagger a_i a_j; a_0^\dagger \gg \\
& + 2\sum_i{}' V_{i0i0} \ll a_i^\dagger a_i a_0; a_0^\dagger \gg + 2\sum_{i,j\neq i}{}' V_{i0j0} \ll a_i^\dagger a_j a_0; a_0^\dagger \gg \\
& + \sum_i{}' V_{i0ii} \ll a_i^\dagger a_i a_i; a_0^\dagger \gg + \sum_{i,j\neq i}{}' V_{i0ij} \ll a_i^\dagger a_i a_j; a_0^\dagger \gg \\
& + \sum_{i,k\neq i}{}' V_{k0ii} \ll a_k^\dagger a_i a_i; a_0^\dagger \gg + \sum_{i,j=k\neq i}{}' V_{j0ij} \ll a_j^\dagger a_i a_j; a_0^\dagger \gg \\
& + \sum_{i,j\neq k\neq i}{}' V_{k0ij} \ll a_k^\dagger a_i a_j; a_0^\dagger \gg
\end{aligned} \tag{13}$$

and

$$EF_{00} = -\epsilon_0 \ll a_0^\dagger; a_0^\dagger \gg -V_0 \ll a_0^\dagger a_0^\dagger a_0; a_0^\dagger \gg - \sum_i{}' \epsilon_{i0} \ll a_i^\dagger; a_0^\dagger \gg$$

$$- 2\sum_i{}' V_{i000} \ll a_i^\dagger a_0^\dagger a_0; a_0^\dagger \gg - \sum_i{}' V_{00i0} \ll a_0^\dagger a_0^\dagger a_i; a_0^\dagger \gg$$

$$- \sum_i{}' V_{ii00} \ll a_i^\dagger a_i^\dagger a_0; a_0^\dagger \gg - \sum_{i,j\neq i}{}' V_{ij00} \ll a_i^\dagger a_j^\dagger a_0; a_0^\dagger \gg$$

$$- 2\sum_i{}' V_{i0i0} \ll a_i^\dagger a_0^\dagger a_i; a_0^\dagger \gg -2 \sum_{i,j\neq i}{}' V_{j0i0} \ll a_j^\dagger a_0^\dagger a_i; a_0^\dagger \gg$$

$$- \sum_i{}' V_{iii0} \ll a_i^\dagger a_i^\dagger a_i; a_0^\dagger \gg - \sum_{i,j\neq i}{}' V_{iji0} \ll a_i^\dagger a_j^\dagger a_i; a_0^\dagger \gg$$

$$- \sum_{i,k\neq i}{}' V_{iik0} \ll a_i^\dagger a_i^\dagger a_k; a_0^\dagger \gg - \sum_{i,j=k\neq i}{}' V_{ijj0} \ll a_i^\dagger a_j^\dagger a_j; a_0^\dagger \gg$$

$$- \sum_{i,j\neq k\neq i}{}' V_{ijk0} \ll a_i^\dagger a_j^\dagger a_k; a_0^\dagger \gg . \tag{14}$$

Thus from the equations for G_{00} and F_{00} we see the interplay between the ground state and the excited states. From Eqs. (13) and (14) we get two non-linearly coupled equations for G_{00} and F_{00} in terms of G_{00}, F_{00} and higher order terms whose equations of motion are further complicated. Thus we get an infinite chain of equations, which is tedious to solve and therefore one normally looks for sophisticated methods to linearize or decouple the two equations and this is done by introducing suitable approximations. One such approximation to decouple the equations is the Wick's decoupling - a truncation process by which the higher order Green functions are all reduced to the lower order Green functions. To do this we put:
$\ll a_0^\dagger a_0 a_0; a_0^\dagger \gg = N_0 G_{00}$, $\ll a_0^\dagger a_0^\dagger a_0; a_0^\dagger \gg = N_0 F_{00}$, $\ll a_i^\dagger a_0 a_0; a_0^\dagger \gg = A_{00} F_{i0}$, $\ll a_0^\dagger a_0^\dagger a_i; a_0^\dagger \gg = A_{00} G_{i0}$, where $A_{ij} = \langle a_i a_j \rangle = \langle a_i^\dagger a_j^\dagger \rangle$ and $N_i = \langle a_i^\dagger a_i \rangle$. Terms like A_{ij} are called anomalous averages, which represent the passage of pairs of atoms from the condensate into excited states and the reverse process.

As a first order correction to the ground state Green functions, we restrict ourselves only to the first excited state. For this we consider only the contributions from G_{10} and F_{10}. Thus we get:

$$(E-\tilde{\epsilon_0})\,G_{00} = \frac{1}{2\pi} + \left(\sum_i{}' V_{00ii}A_{ii} + \sum_{i,j\neq i}{}' V_{00ij}A_{ij}\right) F_{00}$$
$$+\left(V_{1000}A_{00} + 2\sum_{j\neq 1}{}' V_{10j0}A_{j0} + \sum_{i\neq 1}{}' V_{10ii}A_{ii} + \sum_{i,j\neq 1}{}' V_{10ij}A_{ij}\right) F_{10}$$
$$+\left(\epsilon_{01} + 2V_{0010}N_0 + V_{1011}N_1 + 2\sum_{j\neq 1}{}' V_{j01j}N_j\right) G_{10} \qquad (15)$$

$$(E+\tilde{\epsilon_0})\,F_{00} = -\left(\sum_i{}' V_{ii00}A_{ii} + \sum_{i,j\neq i}{}' V_{ij00}A_{ij}\right) G_{00}$$
$$-\left(V_{0010}A_{00} + 2\sum_{j\neq 1}{}' V_{j010}A_{j0} + \sum_{i\neq 1}{}' V_{ii10}A_{ii} + \sum_{i,j\neq 1}{}' V_{ij10}A_{ij}\right) G_{10}$$
$$-\left(\epsilon_{10} + 2V_{1000}N_0 + V_{1110}N_1 + 2\sum_{j\neq 1}{}' V_{1jj0}N_j\right) F_{10} \qquad (16)$$

where $\tilde{\epsilon_0} = \epsilon_0 + V_0N_0 + 2\sum_i' V_{i0i0}N_i$

$$(E-\tilde{\epsilon_{11}})\,G_{10} = \left(V_{1100}A_{00} + 2\sum_{k\neq 1}{}' V_{11k0}A_{k0} + \sum_{k\neq 1}{}' V_{11kk}A_{kk} + \sum_{k,l\neq 1}{}' V_{11kl}A_{kl}\right) F_{10}$$
$$+\left(V_{1011}A_{11} + 2\sum_{j\neq 1}{}' V_{101j}A_{1j} + \sum_{i\neq 1}{}' V_{10ii}A_{ii} + \sum_{i,j\neq 1}{}' V_{10ij}A_{ij}\right) F_{00}$$
$$+\left(\epsilon_{10} + V_{1000}N_0 + 2V_{1110}N_1 + 2\sum_{j\neq 1}{}' V_{1jj0}N_j\right) G_{00} \qquad (17)$$

where $\tilde{\epsilon_{11}} = 2\epsilon_{11} + 2V_{1010}N_0 + V_{1111}N_1 + 2\sum_{j\neq 1}' V_{1jj1}N_j$

$$(E+\tilde{\epsilon_{11}})\,F_{10} = -\left(V_{0011}A_{00} + 2\sum_{k\neq 1}{}' V_{k011}A_{k0} + \sum_{k\neq 1}{}' V_{kk11}A_{kk} + \sum_{k,l\neq 1}{}' V_{kl11}A_{kl}\right) G_{10}$$
$$-\left(V_{1110}A_{11} + 2\sum_{j\neq 1}{}' V_{1j10}A_{1j} + \sum_{i\neq 1}{}' V_{ii10}A_{ii} + \sum_{i,j\neq 1}{}' V_{ij10}A_{ij}\right) G_{00}$$
$$-\left(\epsilon_{01} + V_{0010}N_0 + 2V_{1011}N_1 + 2\sum_{j\neq 1}{}' V_{j01j}N_j\right) F_{00}. \qquad (18)$$

4.1. *Solution in lowest order*

To find G_{00} and F_{00} in the lowest order, we drop the contributions from G_{10} and F_{10} and put:

$$X_1 = \sum_i{}' V_{00ii} A_{ii} + \sum_{i,j\neq i}{}' V_{00ij} A_{ij} = \sum_i{}' V_{ii00} A_{ii} + \sum_{i,j\neq i}{}' V_{ij00} A_{ij} = \sum_{i,j}{}' V_{ij00} A_{ij}.$$

The term $\epsilon_0 + V_0 N_0$ in $\tilde{\epsilon_0}$ is the mean-field Hartree term, arising purely due to the atoms present in the lowest state of the confining potential. The term $2\sum_i' V_{i0i0} N_i$ in the same, is the contribution from the excited states to the energy of the condensed state, arising as a result of the two-body scattering between the condensed and the non-condensed atoms, beyond the usual mean-field theories. Now, the term X_1 which contains the anomalous correlation functions, corresponds to the spectral gap. The appearance of a gap in the energy spectrum of the lowest excitations is a feature of the pair Hamiltonian model as reported earlier by several authors.[32,34,36] Thus we get:

$$(E - \tilde{\epsilon_0})\, G^0_{00} = \frac{1}{2\pi} + X_1 F^0_{00} \tag{19}$$

$$(E + \tilde{\epsilon_0})\, F^0_{00} = -X_1 G^0_{00}. \tag{20}$$

Solving the above two equations we get:

$$G^0_{00}(E) = \frac{1}{2\pi} \frac{E + \tilde{\epsilon_0}}{\left(E^2 - \tilde{\epsilon_0}^2 + X_1^2\right)} \tag{21}$$

$$F^0_{00}(E) = -\frac{1}{2\pi} \frac{X_1}{\left(E^2 - \tilde{\epsilon_0}^2 + X_1^2\right)}. \tag{22}$$

Using Eq. (21) one can find the retarded and advanced propagators by treating the energy E as a complex quantity and these propagators are related to the spectral distribution function $J(\omega)$ through the relation[33,34]:

$$G^0_{00}(\omega + i\eta) - G^0_{00}(\omega - i\eta) = -i(e^{\beta\omega} - 1) J_{00}(\omega) \tag{23}$$

from which we get:

$$J_{00}(\omega) = \frac{1}{2}\left[\left(1 + \frac{\tilde{\epsilon_0}}{\overline{\epsilon}}\right)\frac{\delta(\omega - \overline{\epsilon})}{e^{\beta\omega} - 1} + \left(1 - \frac{\tilde{\epsilon_0}}{\overline{\epsilon}}\right)\frac{\delta(\omega + \overline{\epsilon})}{e^{\beta\omega} - 1}\right]. \tag{24}$$

The average occupation number of the condensed state N_0 is given by:

$$N_0 = \int_{-\infty}^{\infty} J_{00}(\omega)\, d\omega \tag{25}$$

which gives:

$$N_0 = \frac{N}{2}\left[\frac{\tilde{\epsilon_0}}{\overline{\epsilon}} \coth\frac{\beta\overline{\epsilon}}{2} - 1\right]. \tag{26}$$

The change in the condensate number depends on the magnitude and sign of the two-body interactions. Case i) when V_0, the two-body interaction energy in the ground state of the external potential is positive (repulsive interaction) and when the term V_{1010} in $\tilde{\epsilon_0}$ is positive, the energy per condensed boson is increased and as a result the condensate gets depleted. Case ii) when V_{1010} is negative, the energy is reduced and consequently depletion reduces too. When V_0 is negative (attractive interaction), Bose condensation is possible, provided $\tilde{\epsilon_0}$ is still $> X_1$ and depletion depends on the magnitude of $\bar{\epsilon}$.

We now use Eq. (26), to estimate the condensate number for temperatures close to T_c. We consider 10^4 rubidium atoms confined by an isotropic harmonic potential of frequency $\omega = 10^3 s^{-1}$. The transition temperature T_c for this system is ≈ 150 nK. For dilute gases, which satisfy the condition $na^3 \ll 1$, where a is the s-wave scattering length, we make a simple estimate for V_{1010} and V_{1100}, each equals to $V_0/8$. With $N_0 = 95\% N$ $N_1 = 5\% N$, $X_1 \approx 0.001\tilde{\epsilon_0}$, we find that as we move away from the mean field regime, $N_0 \approx 530$, for a temperature of 125 nK. These values may change depending upon the interaction which depends on the states.

4.2. *Higher order correction*

Incorporating Green functions G_{10} and F_{10} that connect the ground state to the first excited state and solving we get the ground state Green function involving higher orders which is:

$$G_{00}^{1} = \frac{1}{2\pi}\left(\frac{(E+\tilde{\epsilon_0})\left(E^2-\tilde{\epsilon_{11}}^2+X_4^2\right)}{\left(E^2-\tilde{\epsilon_0}^2+X_1^2\right)\left(E^2-\tilde{\epsilon_{11}}^2+X_4^2\right)+\left(X_2^2-\tilde{\epsilon_{10}}^2\right)\left(X_3^2-{\epsilon'_{10}}^2\right)}\right) \tag{27}$$

which can be rewritten as:

$$\frac{1}{G_{00}^{1}} = \frac{1}{G_{00}^{0}} + 2\pi\frac{\left(X_2^2-\tilde{\epsilon_{10}}^2\right)\left(X_3^2-{\epsilon'_{10}}^2\right)}{(E+\tilde{\epsilon_0})\left(E^2-\tilde{\epsilon_{11}}^2+X_4^2\right)} \tag{28}$$

where

$$\begin{aligned}
X_2 &= V_{0010}A_{00} + 2\textstyle\sum'_{j\neq 1} V_{j010}A_{j0} + \sum'_{i\neq 1} V_{ii10}A_{ii} + \sum'_{i,j\neq 1} V_{ij10}A_{ij}\\
\tilde{\epsilon_{10}} &= \epsilon_{10} + 2V_{1000}N_0 + V_{1110}N_1 + 2\textstyle\sum'_{j\neq 1} V_{1jj0}N_j\\
X_3 &= V_{1110}A_{11} + 2\textstyle\sum'_{j\neq 1} V_{1j10}A_{1j} + \sum'_{i\neq 1} V_{ii10}A_{ii} + \sum'_{i,j\neq 1} V_{ij10}A_{ij}\\
\epsilon'_{10} &= \epsilon_{01} + V_{0010}N_0 + 2V_{1011}N_1 + 2\textstyle\sum'_{j\neq 1} V_{j01j}N_j\\
X_4 &= V_{0011}A_{00} + 2\textstyle\sum'_{k\neq 1} V_{k011}A_{k0} + \sum'_{k\neq 1} V_{kk11}A_{kk} + \sum'_{k,l\neq 1} V_{kl11}A_{kl}.
\end{aligned}$$

The second term on the right hand side of Eq. (28) is the correction term which arises due to the contribution from the first excited state to the condensed state Green function. The magnitude of this term depends on the matrix elements (which can be either real or complex in general) of the pair-wise scattering and its sign determines whether the condensate number is enhanced or depleted. These matrix elements are not yet known, but in principle, they can be calculated from basic scattering theory.

The equation for the single particle Green function, in principle, contains all the effects of interaction in a many-body system. Due to the macroscopic occupation of the ground state Eq. (26), the density distribution as a function of energy, has a large peak of finite width (in a simple case, it is a Lorentzian). The shift and the width of the peak is determined by the strength of the two-body interactions which deplete the condensate, when the interaction is repulsive. In addition to this, thermal effects cause further depletion of the condensate. As a result, particles are raised to the excited states with the density distribution getting reduced as we move towards the higher excited states. For a confined system, the peaks are of finite but reduced height and width as we advance through the energy levels, and are centred at energies given by the poles of the corresponding Green functions. This is shown as an illustration in Fig. 1. In the quasi-continuum limit when the thermal energy kT is very much larger than the spacing between the energy levels, the discrete peaks vanish and a smooth curve passes through the heights of the peaks.

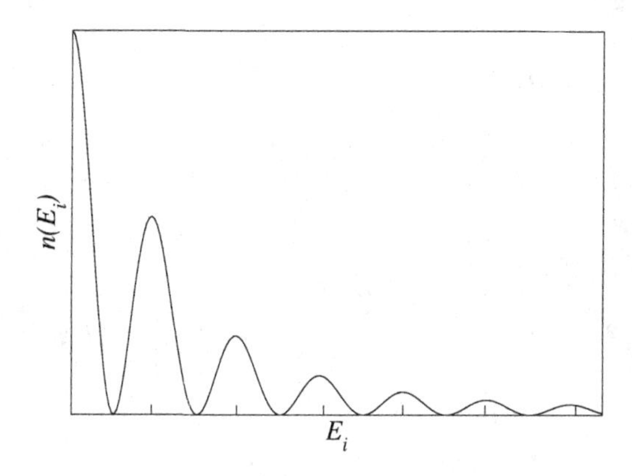

Fig. 1. Density as a function of energy.

5. Specific Heat of Trapped Interacting Bose Gas in the Quasi-Continuum Limit

We now discuss the behaviour of the specific heat of a trapped interacting Bose gas, which is still a subject of interest due to its peculiar features at around the transition temperature. How the interparticle interactions and boundary conditions affect specific heat at around the transition point is very crucial in understanding the nature of the phase transition. In the earlier treatment on Bose systems[37] an expression for the specific heat and transition temperature was obtained for a system of non-interacting bosons. This treatment was later extended to confined ideal bosons in the thermodynamic limit, in which, it has been shown[38–40] that for an ideal Bose gas confined by a 3D harmonic potential, the specific heat is discontinuous at $T = T_c$. This feature exhibits a strong resemblance to that of liquid ^4He

at the "λ"-point. However in the latter case the "λ" transition is attributed to the interparticle interactions.

The effect of interaction in a homogeneous (i.e. not confined by an external potential) system of bosons was studied[32] using a pair Hamiltonian model, in which, interaction was found to introduce a discontinuity at T_c. In the present study, we consider an interacting Bose gas confined by a weak harmonic potential. When the potential is very shallow, the discrete energy levels can then be replaced by a continuous density of states. In order to achieve this, we take the thermodynamic limit, in which we let the range parameter R in the harmonic potential to increase[41,42] (i.e. we let the potential to weaken: $\omega \to 0$) as N, the total number of particles increases. This thermodynamic limit requires the condition that the thermal energy kT is very much larger than the spacing between the energy levels, so that a continuous density of states is justified. As a result of weakening of the potential, the particles in the centre of the trap may be considered as nearly uniform. Therefore, in this quasi-continuum limit our results will coincide with the results given in Ref. [32]. Thus, the expression for the specific heat of an interacting Bose gas, as an explicit function of temperature, for $T > T_c$ is:

$$C_v \propto \left[\frac{15}{4} g_{\frac{5}{2}} (T - T_c)^{3/2} + 3\frac{\epsilon(0)}{k} g_{\frac{3}{2}} (T - T_c)^{1/2} + \left(\frac{\epsilon(0)}{k}\right)^2 g_{\frac{1}{2}} (T - T_c)^{-1/2}\right] \quad (29)$$

where $\epsilon(0)$ is the energy of the $\mathbf{k} = 0$ state and it is related to both the chemical potential μ and the interaction term $v(0)$; $g_{\frac{5}{2}}$, $g_{\frac{3}{2}}$ and $g_{\frac{1}{2}}$ are called Bose integrals which are defined as $g_n(z) = \frac{1}{\Gamma(n)} \int_0^\infty \frac{x^{n-1}}{z^{-1}e^x - 1}$ with $\Gamma(n) = \int_0^\infty e^{-x} x^{n-1} dx$.

Eq. (29) has been derived from the expression for the thermodynamic potential given in Ref. [32]. For temperatures very close to and above T_c, the first two terms could be neglected, the term $(T - T_c)^{-1/2}$ dominates highly and at $T = T_c$ there is a square-root singularity. This singularity has been shown in an earlier work[43] for both the Hartree-Fock and Bogoliubov models of a system of bosons interacting via two-body potentials. Specific heat of an interacting Bose gas, obtained using the pressure equation[32] and expressed in a different form for $T > T_c$ is:

$$\frac{C_v}{Nk} = \left(\frac{15}{4}\frac{1}{\rho\Omega_T} g_{5/2} - \frac{9}{4}\frac{g_{3/2}}{g_{1/2}}\right)\left[1 - \frac{3}{2}\rho v(0)(k_c^2/2m^*)^{-1}\right] \quad (30)$$

where Ω_T and k_c are coherent volume and cut-off wavevector respectively. From the above equation we see that specific heat of an interacting Bose gas is reduced compared to that of an ideal gas by the term $\frac{3}{2}\rho v(0)(k_c^2/2m^*)^{-1}$. In the absence of interaction $v(0) = 0$ and the expression for the specific heat reduces to that of an ideal gas.[37]

It has been shown[32] that for an interacting Bose gas the transition temperature is shifted downwards to T_c, with respect to that of an ideal gas, as shown in Fig. 2. As a result the specific heat above T_c for the former is reduced, in accordance to Eq. (30). However, in both cases, above the transition temperature, the specific

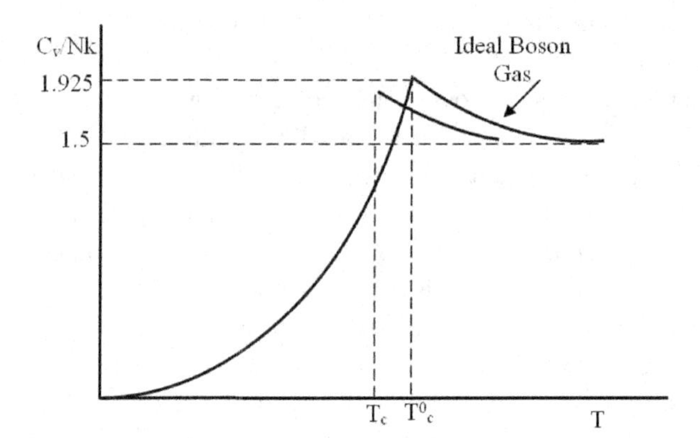

Fig. 2. Specific heat behaviour of the interacting and that of the ideal Bose gas.

heat follows the same trend. We mention in passing that the explicit Eqs. (29) and (30) for the specific heat was not given in Ref. [32].

For an ideal gas, the specific heat exhibits a cusp at T_c when the system is homogeneous and a discontinuity when the system is confined, respectively. When interparticle interactions are turned on, the specific heat exhibits a discontinuity at T_c. For a trapped interacting gas, in the thermodynamic limit, thus we expect a discontinuity at the transition point, due to both confinement and interaction. The nature of this discontinuity depends on the strengths of both the interaction and the confining potential. Further to this, we make a note that interaction reduces the transition temperature with respect to that of an ideal gas. If we take into account the finite-size effects,[38,44] then we can expect a further downward shift in the transition temperature.

6. Conclusion

The main objectives of this paper were to set up a mathematical procedure to calculate the condensate number, using a pair Hamiltonian model and to show the interplay of the condensate with the non-condensate components. This was achieved by setting up a finite-temperature Green function formalism where we have suitably defined the statistical averages and the Heisenberg equations of motion. Due to the interaction terms present in the Hamiltonian we get a hierarchy of never ending equations for the Green functions, for which there is no exact solution. In order to find the exact solution we have truncated the equations using Wick's decoupling. As a result of this physically motivated approximation, we have obtained an expression for the condensate number using the normal ordering Green function in the lowest order. Using this expression, depletion of bosons from the ground state was qualitatively discussed. Introducing the next higher order correction to the ground state Green function, we correspondingly see that depletion would be enhanced or lowered depending upon the magnitude and sign of the matrix elements. These matrix elements are not known yet, however, in principle, they can be calculated.

Finally we have discussed the behaviour of specific heat of the interacting gas in the quasi-continuum limit. Following which, we see how the interparticle interactions and boundary conditions play significant roles in affecting the nature of the specific heat around the transition temperature.

References

1. M. H. Anderson, J. R. Ensher, M. R. Matthews, C. E. Wieman and E. A. Cornell, *Science* **269**, p. 198 (1995).
2. K. B. Davis, M. O. Mewes, M. R. Andrews, N. J. van Druten, D. S. Durfee, D. M. Kurn and W. Ketterle, *Phys. Rev. Lett.* **75**, p. 3969 (1995).
3. C. M. Savage and M. P. Das (eds.), *Bose-Einstein Condensation: from Atomic Physics to Quantum Liquids* (World Scientific, Singapore, 2000).
4. A. S. Parkins and D. F. Walls, *Phys. Rep.* **303**, p. 1 (1998).
5. F. Dalfovo, S. Giorgini, L. P. Pitaevskii and S. Stringari, *Rev. Mod. Phys.* **71**, 463(Apr 1999).
6. A. J. Leggett, *Rev. Mod. Phys.* **73**, p. 307 (2001).
7. A. J. Leggett, *Quantum Liquids: Bose Condensation and Cooper Pairing in Condensed-Matter Systems* (Oxford University Press, Oxford, 2006).
8. E. P. Gross, *Nuovo Cimento* **20**, p. 454 (1963).
9. L. P. Pitaevskii, *Sov. Phys.-JETP* **13**, p. 451 (1961).
10. F. Dalfovo and S. Stringari, *Phys. Rev. A* **53**, p. 2477 (1996).
11. M. Edwards, R. J. Dodd, C. W. Clark, P. A. Ruprecht and K. Burnett, *Phys. Rev. A* **53**, p. R1950 (1996).
12. N. N. Bogoliubov, *J. Phys. (USSR)* **11**, p. 23 (1947).
13. T. D. Lee, K. Huang and C. N. Yang, *Phys. Rev.* **106**, p. 1135 (1957).
14. S. T. Beliaev, *J. Exptl. Theoret. Phy. U.S.S.R.* **34**, p. 417 (1958).
15. N. M. Hugenholtz and D. Pines, *Phys. Rev.* **116**, 489(Nov 1959).
16. S. L. Cornish, *Phys. Rev. Lett.* **85**, p. 1795 (2000).
17. A. Griffin, *Excitations in a Bose-Condensed Liquid* (Cambridge University Press, Cambridge, 1993).
18. P. C. Hohenberg and P. C. Martin, *Ann. Phys. (New York)* **34**, p. 291 (1965).
19. V. N. Popov and L. D. Fadeev, *Sov. Phys.-JETP* **20**, p. 890 (1965).
20. T. H. Cheung and A. Griffin, *Phys. Rev. A* **4**, p. 237 (1971).
21. P. O. Fedichev and G. V. Shlyapnikov, *Phys. Rev. A* **58**, p. 3146 (1998).
22. M. Rusch and K. Burnett, *Phys. Rev. A* **59**, p. 3851 (1999).
23. S. A. Morgan, *J. Phys. B: At. Mol. Opt. Phys.* **33**, p. 3847 (2000).
24. S. A. Morgan, *Phys. Rev. A* **69**, p. 023609 (2004).
25. V. L. Bonch-Bruevich and S. V. Tyablikov, *The Green Function Method in Statistical Mechanics* (Interscience, Amsterdam, North Holland, 1962).
26. A. I. Alekseev, *Sov. Phys. Usp.* **4**, p. 23 (1961).
27. L. P. Kadanoff and G. Baym, *Quantum Statistical Mechanics* (Benjamin, New York, 1962).
28. P. C. Martin and J. Schwinger, *Phys. Rev.* **115**, p. 1342 (1959).
29. A. L. Fetter and J. D. Walecka, *Quantum Theory of Many-Particle Systems* (McGraw-Hill, New York, 1971).
30. A. A. Abrikosov, L. P. Gorkov and I. E. Dzyaloshinski, *Methods of Quantum Field Theory in Statistical Physics* (Prentice-Hall, Englewood Cliffs, NJ, 1963).
31. A. Griffin, *Phys. Rep.* **304**, p. 1 (1988).
32. M. Luban, *Phys. Rev.* **128**, 965(Oct 1962).

33. D. N. Zubarev, *Sov. Phys. Usp.* **3**, p. 320 (1960).
34. D. ter Haar, *Lectures on Selected Topics in Statistical Mechanics* (Pergamon Press, Oxford, 1977).
35. J. W. Halley (ed.), *Correlation Functions and Quasiparticle Interactions in Condensed Matter* (Plenum Press, New York, 1978).
36. M. Girardeau and R. Arnowitt, *Phys. Rev.* **113**, p. 755 (1959).
37. K. Huang, *Statistical Mechanics* (Wiley, New York, 1987).
38. C. J. Pethick and H. Smith, *Bose-Einstein Condensation in Dilute Gases* (Cambridge University Press, Cambridge, 2002).
39. S. Grossman and M. Holthaus, *Phys. Lett. A* **208**, p. 188 (1995), and references therein.
40. S. Grossman and M. Holthaus, *Z. Naturforsch* **50a**, p. 921 (1995).
41. R. Masut and W. J. Mullin, *Am. J. Phys.* **47**, p. 493 (1979).
42. W. J. Mullin, *J. Low. Temp. Phys.* **106**, p. 615 (1997).
43. M. Luban and W. D. Grobman, *Phys. Rev. Lett.* **17**, 182(Jul 1966).
44. S. H. Kim, G. Gnanapragasam and M. P. Das, submitted.

VARIATION OF THE FUNDAMENTAL CONSTANTS: THEORY AND OBSERVATIONS

V. V. FLAMBAUM*

School of Physics, University of New South Wales,
Sydney, 2052 Australia
and
Institute for Advanced Study, Massey University (Albany Campus),
Private Bag 102904, North Shore MSC Auckland, New Zealand
**E-mail: flambaum@phys.unsw.edu.au*

Review of recent works devoted to the variation of the fine structure constant α, strong interaction and fundamental masses (Higgs vacuum) is presented. The results from Big Bang nucleosynthesis, quasar absorption spectra, and Oklo natural nuclear reactor data give us the space-time variation on the Universe lifetime scale. Comparison of different atomic clocks gives us the present time variation. Assuming linear variation with time we can compare different results. The best limit on the variation of the electron-to-proton mass ratio $\mu = m_e/M_p$ and $X_e = m_e/\Lambda_{QCD}$ follows from the quasar absorption spectra:[1] $\dot{\mu}/\mu = \dot{X}_e/X_e = (1 \pm 3) \times 10^{-16}$ yr^{-1}. A combination of this result and the atomic clock results[2,3] gives the best limt on variation of α: $\dot{\alpha}/\alpha = (-0.8 \pm 0.8) \times 10^{-16}$ yr^{-1}. The Oklo natural reactor gives the best limit on the variation of $X_s = m_s/\Lambda_{QCD}$ where m_s is the strange quark mass:[4,5] $|\dot{X}_s/X_s| < 10^{-18}$ yr^{-1}. Note that the Oklo data can not give us any limit on the variation of α since the effect of α there is much smaller than the effect of X_s and should be neglected.

Huge enhancement of the relative variation effects happens in transitions between close atomic, molecular and nuclear energy levels. We suggest several new cases where the levels are very narrow. Large enhancement of the variation effects is also possible in cold atomic and molecular collisions near Feshbach resonance.

How changing physical constants and violation of local position invariance may occur? Light scalar fields very naturally appear in modern cosmological models, affecting parameters of the Standard Model (e.g. α). Cosmological variations of these scalar fields should occur because of drastic changes of matter composition in Universe: the latest such event is rather recent (about 5 billion years ago), from matter to dark energy domination. Massive bodies (stars or galaxies) can also affect physical constants. They have large scalar charge S proportional to number of particles which produces a Coulomb-like scalar field $U = S/r$. This leads to a variation of the fundamental constants proportional to the gravitational potential, e.g. $\delta\alpha/\alpha = k_\alpha \delta(GM/rc^2)$. We compare different manifestations of this effect. The strongest limits[6] $k_\alpha + 0.17k_e = (-3.5 \pm 6) \times 10^{-7}$ and $k_\alpha + 0.13k_q = (-1 \pm 17) \times 10^{-7}$ are obtained from the measurements of dependence of atomic frequencies on the distance from Sun[2,7] (the distance varies due to the ellipticity of the Earth's orbit).

Keywords: Fundamental constants; cosmological variation; quasar spectra; atomic clocks; unification theories.

1. Introduction

A search for the variations of the fundamental constants is currently a very popular research topic. Theories unifying gravity and other interactions suggest the possibility of spatial and temporal variation of physical "constants" in the Universe (see, e.g. Refs. 8 and 9). Moreover, there exists a mechanism for making all coupling constants and masses of elementary particles both space and time dependent, and influenced by local circumstances (see e.g. review[9]). The variation of coupling constants can be non-monotonic (for example, damped oscillations).

These variations are usually associated with the effect of massless (or very light) scalar fields. One candidate is the dilaton: a scalar which appears in string theories together with a graviton, in a massless multiplet of closed string excitations. Other scalars naturally appear in cosmological models, in which our Universe is a "brane" floating in a space of larger dimensions. The scalars are simply brane coordinates in extra dimensions. However, the only relevant scalar field recently discovered, the cosmological dark energy, so far does not show visible variations. Available observational limits on physical constant variations at present time are quite strict, allowing only scalar coupling tiny in comparison with gravity.

A possible explanation was suggested by Damour et al.[10,11] who pointed out that cosmological evolution of scalars naturally leads to their self-decoupling. Damour and Polyakov have further suggested that variations should happen when the scalars get excited by some physical change in the Universe, such as the phase transitions or other drastic change in the equation of State of the Universe. They considered few of them, but since the time of their paper a new fascinating transition has been discovered: from matter dominated (decelerating) era to dark energy dominated (accelerating) era. It is relatively recent event, corresponding to cosmological redshift $z \approx 0.5$.

The time dependence of the perturbation related to it can be calculated, and it turned out[13,14] that the self-decoupling process is effective enough to explain why after this transition the variation of constants is as small as observed in laboratory experiments at the present time, as well as at Oklo ($\sim$ 2 billion years ago or $z = 0.14$) and isotopes ratios in meteorites (4.6 billion years to now, $z = 0.45 - 0$), while being at the same time consistent with possible observations of the variations of the electromagnetic fine structure constant at $z \sim 1$.

Another topic we will address here is similar variations of constants in space, near massive bodies such as stars (Sun), pulsars, Galaxy. We will compare possible sensitivities related with different possible objects, point out limitations following from some recent experiments with atomic clocks and suggest new measurements (this part is based on Ref. 6).

Recent observations have produced several hints for the variation of the fine structure constant, $\alpha = e^2/\hbar c$, strength constant of the strong interaction and masses in Big Bang nucleosynthesis, quasar absorption spectra and Oklo natural nuclear reactor data (see e.g. Refs. 15–17, 19). However, a majority of publications report only limits on possible variations (see e.g. reviews[9,20]). A very sensitive method to study the variation in a laboratory consists of the comparison of different optical and microwave atomic clocks (see recent measurements in Refs. 2, 21–28).

Sensitivity to temporal variation of the fundamental constants may be strongly enhanced in transitions between narrow close levels of different nature. Huge enhancement of the relative variation effects can be obtained in transition between the almost degenerate levels in atoms,[29–33] molecules[1,34–37] and nuclei.[38,39]

2. Optical Spectra

2.1. *Comparison of quasar absorption spectra with laboratory spectra*

To perform measurements of α variation by comparison of cosmic and laboratory optical spectra we developed a new approach[29,40] which improves the sensitivity to a variation of α by more than an order of magnitude. The relative value of any relativistic corrections to atomic transition frequencies is proportional to α^2. These corrections can exceed the fine structure interval between the excited levels by an order of magnitude (for example, an s-wave electron does not have the spin-orbit splitting but it has the maximal relativistic correction to energy). The relativistic corrections vary very strongly from atom to atom and can have opposite signs in different transitions (for example, in s-p and d-p transitions). Thus, any variation of α could be revealed by comparing different transitions in different atoms in cosmic and laboratory spectra.

This method provides an order of magnitude precision gain compared to measurements of the fine structure interval. Relativistic many-body calculations are used to reveal the dependence of atomic frequencies on α for a range of atomic species observed in quasar absorption spectra.[29,30,40,41] It is convenient to present results for the transition frequencies as functions of α^2 in the form

$$\omega = \omega_0 + qx, \tag{1}$$

where $x = (\frac{\alpha}{\alpha_0})^2 - 1 \approx \frac{2\delta\alpha}{\alpha}$ and ω_0 is a laboratory frequency of a particular transition. We stress that the second term contributes only if α deviates from the laboratory value α_0. We performed accurate many-body calculations of the coefficients q for all transtions of astrophysical interest (strong E1 transtions from the ground state) in Mg, Mg II, Fe II, Cr II, Ni II, Al II, Al III, Si II, and Zn II. It is very important that this set of transtions contains three large classes: positive shifters (large positive coefficients $q > 1000$ cm^{-1}), negative shifters (large negative coefficients $q < -1000$ cm^{-1}) and anchor lines with small values of q. This gives us an excellent control of systematic errors since systematic effects do not "know" about sign and magnitude of q. Comparison of cosmic frequencies ω and laboratory frequencies ω_0 allows us to measure $\frac{\delta\alpha}{\alpha}$.

Three independent samples of data contaning 143 absorption systems spread over red shift range $0.2 < z < 4.2$. The fit of the data gives[15] is $\frac{\delta\alpha}{\alpha} = (-0.543 \pm$

$0.116) \times 10^{-5}$. If one assumes the linear dependence of α on time, the fit of the data gives $d \ln \alpha/dt = (6.40 \pm 1.35) \times 10^{-16}$ per year (over time interval about 12 billion years). A very extensive search for possible systematic errors has shown that known systematic effects can not explain the result (It is still not completely excluded that the effect may be imitated by a large change of abundances of isotopes during last 10 billion years. We have checked that different isotopic abundances for any single element can not imitate the observed effect. It may be an improbable "conspiracy" of several elements).

Recently our method and calculations[29,30,40,41] were used by two other groups.[42,43] However, they have not detected any variation of α. Most probably, the difference is explained by some undiscovered systematic effects. However, another explanation is not excluded. These results of Ref. 15 are based on the data from the Keck telescope which is located in the Northen hemisphere (Hawaii). The results of Refs. 42 and 43 are based on the data from the different telescope (VLT) located in the Southern hemisphere (Chile). Therefore, the difference in the results may be explained by the spatial variation of α.

Recently the results of Ref. 42 were questioned in Ref. 44. Re-analysis of Ref. 42 data revealed flawed parameter estimation methods. The authors of Ref. 44 claim that the same spectral data fitted more accurately give $\frac{\delta\alpha}{\alpha} = (-0.44 \pm 0.16) \times 10^{-5}$ (instead of $\frac{\delta\alpha}{\alpha} = (-0.06 \pm 0.06) \times 10^{-5}$ in Ref. 42). However, even this revised result may require further revision.

Using opportunity I would like to ask for new, more accurate laboratory measurements of UV transition frequencies which have been observed in the quasar absorption spectra. The "shopping list" is presented in Ref. 45. We also need the laboratory measurements of isotopic shifts - see Ref. 45. We have performed very complicated calculations of these isotopic shifts.[46] However, the accuracy of these calculations in atoms and ions with open d-shell (like Fe II, Ni II, Cr II, Mn II, Ti II) may be very low. The measurements for at list few lines are needed to test these calculations. These measurements would be very important for a study of evolution of isotope abundances in the Universe, to exclude the systematic effects in the search for α variation and to test models of nuclear reactions in stars and supernovi.

2.2. *Optical clocks*

Optical clocks also include transitions which have positive, negative or small constributions of the relativistic corrections to frequencies. We used the same methods of the relativistic many-body calculations to calculate the dependence on α.[29,30,47] The coefficients q for optical clock transitions may be substantially larger than in cosmic transitions since the clock transitions are often in heavy atoms (Hg II, Yb II, Yb III, etc.) while cosmic spectra contain mostly light atoms lines ($Z < 33$). The relativistic effects are proporitional to $Z^2\alpha^2$.

3. Enhanced Effects of α Variation in Atoms

An enhancement of the relative effect of α variation can be obtained in transition between the almost degenerate levels in Dy atom.[29,30] These levels move in opposite directions if α varies. The relative variation may be presented as $\delta\omega/\omega = K\delta\alpha/\alpha$ where the coefficient K exceeds 10^8. Specific values of $K = 2q/\omega$ are different for different hyperfine components and isotopes which have different ω; $q = 30,000$ cm^{-1}, $\omega \sim 10^{-4}$ cm^{-1}. An experiment is currently underway to place limits on α variation using this transition.[32,33] The current limit is $\dot{\alpha}/\alpha = (-2.7 \pm 2.6) \times 10^{-15}$ yr^{-1}. Unfortunately, one of the levels has quite a large linewidth and this limits the accuracy.

Several enhanced effects of α variation in atoms have been calculated in Ref. 31.

4. Enhanced Effects of α Variation in Molecules

The relative effect of α variation in microwave transitions between very close and narrow rotational-hyperfine levels may be enhanced 2-3 orders of magnitude in diatomic molecules with unpaired electrons like LaS, LaO, LuS, LuO, YbF and similar molecular ions.[35] The enhancement is a result of cancellation between the hyperfine and rotational intervals; $\delta\omega/\omega = K\delta\alpha/\alpha$ where the coefficients K are between 10 and 1000.

This enhancement may also exist in a large number of molecules due to cancelation between the ground state fine structure ω_f and vibrational interval ω_v ($\omega = \omega_f - n\omega_v \approx 0$, $\delta\omega/\omega = K(2\delta\alpha/\alpha - 0.5\delta\mu/\mu)$, $K \gg 1$, $\mu = m_e/M_p$ - see Ref. 37). The intervals between the levels are conveniently located in microwave frequency range and the level widths are very small. Required accuracy of the shift measurements is about 0.01-1 Hz. As examples, we consider molecules Cl_2^+, CuS, IrC, SiBr and HfF^+. An enhancement due to the cancellation between the electron and vibrational intervals in Cs_2 molecule was suggested earlier by D. DeMille.[34]

5. Variation of the Strong Interaction

The hypothetical unification of all interactions implies that a variation in α should be accompanied by a variation of the strong interaction strength and the fundamental masses. For example, the grand unification models discussed in Ref. 8 predicts the quantum chromodynamics (QCD) scale Λ_{QCD} (defined as the position of the Landau pole in the logarithm for the running strong coupling constant, $\alpha_s(r) \sim 1/\ln(\Lambda_{QCD}r/\hbar c))$ is modified as $\delta\Lambda_{QCD}/\Lambda_{QCD} \approx 34\ \delta\alpha/\alpha$. The variations of quark mass m_q and electron masses m_e (related to variation of the Higgs vaccuum field which generates fundamental masses) in this model are given by $\delta m/m \sim 70\ \delta\alpha/\alpha$, giving an estimate of the variation for the dimensionless ratio

$$\frac{\delta(m/\Lambda_{QCD})}{(m/\Lambda_{QCD})} \sim 35\frac{\delta\alpha}{\alpha} \tag{2}$$

The coefficient here is model dependent but large values are generic for grand unification models in which modifications come from high energy scales; they appear because the running strong-coupling constant and Higgs constants (related to mass) run faster than α.

Indeed, the strong (i=3), and electroweak (i=1,2) inverse coupling constants have the following dependence on the scale ν and normalization point ν_0:

$$\alpha_i^{-1}(\nu) = \alpha_i^{-1}(\nu_0) + b_i ln(\nu/\nu_0). \tag{3}$$

In the Standard Model $2\pi b_i = 41/10, -19/6, -7$ and the couplings are related as $\alpha^{-1} = (5/3)\alpha_1^{-1} + \alpha_2^{-1}$. There are two popular scenarios of Grand Unification: with the standard model as well as for its minimal supersymmetric extension (MSSM). In the latter case 3 curves for α_i (i=1,2,3) cross at one point, believed to be a "root" of the three branches (electromagnetic, weak and strong). One may select the unification point for ν_0, and for example, $\nu = m_Z$ is the Z-boson mass (String theories lead to more complicated "trees", which however also have a singly "root", at a string scale Λ_s and bare string coupling g_s.)

Basically there are two possibilities. If one assumes that only $\alpha_{GUT} \equiv \alpha_i(\nu_0)$ varies, the Eq. (3) gives us the same shifts for all inverse couplings

$$\delta\alpha_1^{-1} = \delta\alpha_2^{-1} = \delta\alpha_3^{-1} = \delta\alpha_{GUT}^{-1}. \tag{4}$$

If so, the variation of the strong interaction constant $\alpha_3(m_z)$ is much larger than the variation of the em constant α, $\delta\alpha_3/\alpha_3 = (\alpha_3/\alpha_1)\delta\alpha_1/\alpha_1$.

Another option is the variation of the GUT scale (ν/ν_0 in Eq. (3)). If so, quite different relations between variations of the three coupling follows

$$\delta\alpha_1^{-1}/b_1 = \delta\alpha_2^{-1}/b_2 = \delta\alpha_3^{-1}/b_3. \tag{5}$$

Note that now variations have different sign since the one loop coefficients b_i have different sign for 1 and 2,3. Another unclear issue is the modification of lepton/quark masses, which are proportional to Higgs vacuum expectation value and thus depend on the mechanism of electroweak symmetry breaking.

If these models are correct, the variation in electron or quark masses and the strong interaction scale may be easier to detect than a variation in α. One can only measure the variation of dimensionless quantities. The variation of m_q/Λ_{QCD} can be extracted from consideration of Big Band nucleosynthesis, quasar absorption spectra and the Oklo natural nuclear reactor, which was active about 1.8 billion years ago.[4] There are some hints for the variation in Big Bang Nucleosynthesis ($\sim 10^{-3}$ - see Ref. 16) and Oklo ($\sim 10^{-9}$ - see Ref. 17) data. However, these results are not confirmed by new studies.[5,18]

The results from Oklo natural nuclear reactor are based on the measurement of the position of very low energy resonance ($E_r = 0.1$ eV) in neutron capture by ^{149}Sm nucleus. The estimate of the shift of this resonance induced by the variation of α have been done long time ago in Ref. 48. Recently we performed a rough estimate of the effect of the variation of m_q/Λ_{QCD}.[4] The final result is

$$\delta E_r \approx 10^6 eV(\frac{\delta\alpha}{\alpha} - 10\frac{\delta X_q}{X_q} + 100\frac{\delta X_s}{X_s}) \qquad (6)$$

where $X_q = m_q/\Lambda_{QCD}$, $X_s = m_s/\Lambda_{QCD}$, $m_q = (m_u + m_d)/2$ and m_s is the strange quark mass. Refs. 5 found that $|\delta E_r| < 0.1$ eV. This gives us a limit

$$|0.01\frac{\delta\alpha}{\alpha} - 0.1\frac{\delta X_q}{X_q} + \frac{\delta X_s}{X_s}| < 10^{-9}. \qquad (7)$$

The contribution of the α variation in this equation is very small and should be neglected since the accuracy of the calculation of the main term is low. Thus, the Oklo data can not give any limit on the variation of α. Assuming linear time dependence during last 2 billion years we obtain an estimate $|\dot{X}_s/X_s| < 10^{-18}$ yr^{-1}.

The proton mass is proportional to Λ_{QCD} ($M_p \sim 3\Lambda_{QCD}$), therefore, the measurements of the variation of the electron-to-proton mass ratio $\mu = m_e/M_p$ is equivalent to the measurements of the variation of $X_e = m_e/\Lambda_{QCD}$. Two new results have been obtained recently using quasar absorption spectra. In our paper[49] the varition of the ratio of the hydrogen hyperfine frequency to optical frequencies in ions have been measured. The result is consistent with no variation of $X_e = m_e/\Lambda_{QCD}$. However, in the recent paper[19] the variation was detected at the level of 4 standard deviations: $\frac{\delta X_e}{X_e} = \frac{\delta\mu}{\mu} = (-2.4 \pm 0.6) \times 10^{-5}$. This result is based on the hydrogen molecule spectra. Note, however, that the difference between the zero result of Ref. 49 and non-zero result of Ref. 19 may be explained by a space-time variation of X_e. The variation of X_e in Ref. 19 is substantially larger than the variation of α measured in Refs. 15, 42. This may be considered as an argument in favour of Grand Unification theories of the variation.[8]

Recently we obtained the limit on the space-time variation of the ratio of the proton mass to the electron mass based on comparison of quasar absorption spectra of NH_3 with CO, HCO^+ and HCN rotational spectra.[1] For the inversion transition in NH_3 ($\lambda \approx 1.25$ cm^{-1}) the relative frequency shift is significantly enhanced: $\delta\omega/\omega = 4.46\,\delta\mu/\mu$. This enhancement allows one to increase sensitivity to the variation of μ using NH_3 spectra for high redshift objects. We use published data on microwave spectra of the object B0218+357 to place the limit $\delta\mu/\mu = (-0.6 \pm 1.9) \times 10^{-6}$ at redshift $z = 0.6847$; this limit is several times better than the limits obtained by different methods and may be significantly improved. Assuming linear time dependence we obtain[1] $\dot{\mu}/\mu = \dot{X}_e/X_e = (1 \pm 3) \times 10^{-16}$ yr^{-1}.

6. Microwave Clocks

Karshenboim[50] has pointed out that measurements of ratios of hyperfine structure intervals in different atoms are sensitive to variations in nuclear magnetic moments. However, the magnetic moments are not the fundamental parameters and can not be directly compared with any theory of the variations. Atomic and nuclear calculations are needed for the interpretation of the measurements. We have performed both atomic calculations of α dependence[29,30,47] and nuclear calculations of $X_q = m_q/\Lambda_{QCD}$ dependence[3] for all microwave transitions of current experimental interest including hyperfine transitions in ^{133}Cs, ^{87}Rb, ^{171}Yb$^+$, ^{199}Hg$^+$, ^{111}Cd, ^{129}Xe, ^{139}La, ^{1}H, ^{2}H and ^{3}He. The results for the dependence of the transition frequencies on variation of α, $X_e = m_e/\Lambda_{QCD}$ and $X_q = m_q/\Lambda_{QCD}$ are presented in Ref. 3 (see the final results in the Table IV of Ref. 3). Also, one can find there experimental limits on these variations which follow from the recent measurements. The accuracy is approaching 10^{-15} per year. This may be compared to the sensitivity $\sim 10^{-5} - 10^{-6}$ per 10^{10} years obtained using the quasar absorption spectra.

According to Ref. 3 the frequency ratio Y of the 282-nm ^{199}Hg$^+$ optical clock transition to the ground state hyperfine transition in ^{133}Cs has the following dependence on the fundamental constants:

$$\dot{Y}/Y = -6\dot{\alpha}/\alpha - \dot{\mu}/\mu - 0.01\dot{X}_q/X_q. \tag{8}$$

In Ref. 2 this ratio has been measured: $\dot{Y}/Y = (0.37 \pm 0.39) \times 10^{-15}$ yr^{-1}. Assuming linear time dependence we obtained the quasar result[1] $\dot{\mu}/\mu = \dot{X}_e/X_e = (1 \pm 3) \times 10^{-16}$ yr^{-1}. A combination of this result and the atomic clock result[2] for Y gives the best limt on the variation of α: $\dot{\alpha}/\alpha = (-0.8 \pm 0.8) \times 10^{-16}$ yr^{-1}. Here we neglected the small ($\sim 1\%$) contribution of X_q.

7. Enhanced Effect of Variation of α and Strong Interaction in UV Transition of ^{229}Th Nucleus (Nuclear Clock)

A very narrow level (3.5 ± 1) eV above the ground state exists in ^{229}Th nucleus[51] (in Ref. 52 the energy is (5.5 ± 1) eV, in Ref. 53 the energy is (7.6 ± 0.5) eV). The position of this level was determined from the energy differences of many high-energy γ-transitions (between 25 and 320 KeV) to the ground and excited states. The subtraction produces the large uncertainty in the position of the 3.5 eV excited state. The width of this level is estimated to be about 10^{-4} Hz.[54] This would explain why it is so hard to find the direct radiation in this very weak transition. The direct measurements have only given experimental limits on the width and energy of this transition (see e.g. Ref. 55). A detailed discussion of the measurements (including several unconfirmed claims of the detection of the direct radiation) is presented in Ref. 54. However, the search for the direct radiation continues.[56]

The ^{229}Th transition is very narrow and can be investigated with laser spectroscopy. This makes ^{229}Th a possible reference for an optical clock of very high accuracy, and opens a new possibility for a laboratory search for the varitation of the fundamental constants.[39]

As it is shown in Ref. 38 there is an additional very important advantage. The relative effects of variation of α and m_q/Λ_{QCD} are enhanced by 5 orders of magnitude. A rough estimate for the relative variation of the ^{229}Th transition frequency is

$$\frac{\delta\omega}{\omega} \approx 10^5 (2\frac{\delta\alpha}{\alpha} + 0.5\frac{\delta X_q}{X_q} - 5\frac{\delta X_s}{X_s})\frac{7\,eV}{\omega} \tag{9}$$

where $X_q = m_q/\Lambda_{QCD}$, $X_s = m_s/\Lambda_{QCD}$, $m_q = (m_u + m_d)/2$ and m_s is the strange quark mass. Therefore, the Th experiment would have the potential of improving the sensitivity to temporal variation of the fundamental constants by many orders of magnitude.

Note that there are other narrow low-energy levels in nuclei, e.g. 76 eV level in ^{235}U with the 26.6 minutes lifetime (see e.g. Ref. 39). One may expect a similar enhancement there. Unfortunetely, this level can not be reached with usual lasers. In principle, it may be investigated using a free-electron laser or synchrotron radiation. However, the accuracy of the frequency measurements is much lower in this case.

8. Enhancement of Variation of Fundamental Constants in Ultracold Atom and Molecule Systems Near Feshbach Resonances

Scattering length A, which can be measured in Bose-Einstein condensate and Feshbach molecule experiments, is extremely sensitive to the variation of the electron-to-proton mass ratio $\mu = m_e/m_p$ or $X_e = m_e/\Lambda_{QCD}$:[57]

$$\frac{\delta A}{A} = K\frac{\delta\mu}{\mu} = K\frac{\delta X_e}{X_e}, \tag{10}$$

where K is the enhancement factor. For example, for Cs-Cs collisions we obtained $K \sim 400$. With the Feshbach resonance, however, one is given the flexibility to adjust position of the resonance using external fields. Near a narrow magnetic or an optical Feshbach resonance the enhancement factor K may be increased by many orders of magnitude.

9. Changing Physics Near Massive Bodies

In this section I follow Ref. 6.

The reason gravity is so important at large scales is that its effect is additive. The same should be true for massless (or very light) scalars: its effect near large body is proportional to the number of particles in it.

For not-too-relativistic objects, like the usual stars or planets, both their total mass M and the total scalar charge Q are simply proportional to the number of nucleons in them, and thus the scalar field is simply proportional to the gravitational potential

$$\phi - \phi_0 = \kappa(GM/rc^2)\,. \tag{11}$$

Therefore, we expect that the fundamental constants would also depend on the position via the gravitational potential at the the measurement point.

Naively, one may think that the larger is the dimensionless gravity potential (GM/rc^2) of the object considered, the better. However, different objects allow for quite different accuracy.

Let us mention few possibilities, using as a comparison parameter the product of gravity potential divided by the tentative relative accuracy

$$P = (GM/rc^2)/(accuracy). \tag{12}$$

(i) Gravity potential on Earth is changing due to ellipticity of its orbit: the corresponding variation of the Sun graviational potential is $\delta(GM/rc^2) = 3.3 \cdot 10^{-10}$. The accuracy of atomic clocks in laboratory conditions approaches 10^{-16}, and so $P \sim 3 \cdot 10^6$. However, comparing clocks on Earth and distant satellite one may get variation of the Earth graviational potential $\delta(GM/rc^2) \sim 10^{-9}$ and $P \sim 10^7$. The space mission was recently discussed, e.g. in the proposal Ref. 59 and references therein. Note that the matter composition of Earth and Sun is very different, therefore, the proportionality coefficients κ in Eq (11) may also be different. Indeed, the first example (Sun) is mainly sensitive to the scalar potentials of electrons and protons while the second example (Earth) is in addition sensitive to the scalar potentials of neutrons and virtual mesons mediating the nuclear forces (the nuclear binding energy).

(ii) Sun (or other ordinary stars) has $GM/rc^2 \sim 2 \cdot 10^{-7}$. Assuming accuracy 10^{-8} in the measurements of atomic spectra near the surface we get $P \sim 10$. However, a mission with modern atomic clocks sent to the Sun would have $P \sim 10^8$ or so, see details in the proposal Ref. 58.

(iii) The stars at different positions inside our (or other) Galaxy have gravitational potential difference of the order of 10^{-7}, and (like for the Sun edge) one would expect $P \sim 10$. Clouds which give the observable absorption lines in quasar spectra have also different gravitational potentials (relative to Earth), of comparable magnitude.

(iv) White/brown dwarfs have $GM/rc^2 \sim 3 \cdot 10^{-4}$, and in some cases rather low temperature. We thus get $P \sim 3 \cdot 10^4$.

(v) Neutron stars have very large gravitational potential $GM/rc^2 \sim 0.1$, but high temperature and magnetic fields make accuracy of atomic spectroscopy rather problematic, we give tentative accuracy 1 percent. $P \sim 10$.

(vi) Black holes, in spite of its large gravitational potential, have no scalar field outside the Shwartzschield radius, and thus are not useful for our purpose.

Accuracy of the atomic clocks is so high because they use extremely narrow lines. At this stage, therefore, star spectroscopy seem not to be competitive: the situation may change if narrow lines be identified.

Now let us see what is the best limit available today. As an example we consider recent work[2] who obtained the following value for the half-year variation of the frequency ratio of two atomic clocks: (i) optical transitions in mercury ions $^{199}Hg^+$ and (ii) hyperfine splitting in ^{133}Cs (the frequency standard). The limit obtained is

$$\delta ln(\frac{\omega_{Hg}}{\omega_{Cs}}) = (0.7 \pm 1.2) \cdot 10^{-15}. \tag{13}$$

For Cs/Hg frequency ratio of these clocks the dependence on the fundamental constants was evaluated in Ref. 3 with the result

$$\delta ln(\frac{\omega_{Hg}}{\omega_{Cs}}) = -6\frac{\delta\alpha}{\alpha} - 0.01\frac{\delta(m_q/\Lambda_{QCD})}{(m_q/\Lambda_{QCD})} - \frac{\delta(m_e/M_p)}{(m_e/M_p)}. \tag{14}$$

Another work Ref. 60 compare H and ^{133}Cs hyperfine transitions. The amplitude of the half-year variation found were

$$|\delta ln(\omega_H/\omega_{Cs})| < 7 \cdot 10^{-15}. \tag{15}$$

The sensitivity[3]

$$\delta ln(\frac{\omega_H}{\omega_{Cs}}) = -0.83\frac{\delta\alpha}{\alpha} - 0.11\frac{\delta(m_q/\Lambda_{QCD})}{(m_q/\Lambda_{QCD})}. \tag{16}$$

There is no sensitivity to m_e/M_p because they are both hyperfine transitions.

As motivated above, we assume that scalar and gravitational potentials are proportional to each other, and thus introduce parameters k_i as follows

$$\frac{\delta\alpha}{\alpha} = k_\alpha \delta(\frac{GM}{rc^2}) \tag{17}$$

$$\frac{\delta(m_q/\Lambda_{QCD})}{(m_q/\Lambda_{QCD})} = k_q \delta(\frac{GM}{rc^2}) \tag{18}$$

$$\frac{\delta(m_e/\Lambda_{QCD})}{(m_e/\Lambda_{QCD})} = \frac{\delta(m_e/M_p)}{(m_e/M_p)} = k_e \delta(\frac{GM}{rc^2}) \tag{19}$$

where in the r.h.s. stands half-year variation of Sun's gravitational potential on Earth.

In such terms, the results of Cs/Hg frequency ratio measurement[2] can be rewritten as

$$k_\alpha + 0.17k_e = (-3.5 \pm 6) \cdot 10^{-7}. \tag{20}$$

The results of Cs/H frequency ratio measurement[60] can be presented as

$$|k_\alpha + 0.13k_q| < 2.5 \cdot 10^{-5}. \tag{21}$$

Finally, the result of recent measurement[7] of Cs/H frequency ratio can be presented as

$$k_\alpha + 0.13k_q = (-1 \pm 17) \cdot 10^{-7}. \tag{22}$$

The sensitivity coefficients for other clocks have been discussed above.

Acknowledgments

The author is grateful to E. Shuryak for valuable contribution to the part of this work describing dependence of the fundamental constants on the gravitational potential, and to D. Budker and S. Schiller for useful comments. This work is supported by the Australian Research Council.

References

1. V.V. Flambaum, M.G. Kozlov, *Phys. Rev. Lett.* (in press), arxiv: 0704.2301 astro-ph
2. Fortier *et al.*, *Phys. Rev. Lett.* **98**, 070801, (2007).
3. V.V. Flambaum and A.F. Tedesco, *Phys.Rev. C* **73**, 055501 1-9 (2006).
4. V.V. Flambaum and E.V. Shuryak, *Phys. Rev. D* **65**, 103503 (2002); V.F. Dmitriev and V.V. Flambaum, *Phys. Rev. D* **67**, 063513 (2003); V.V. Flambaum and E.V. Shuryak, *Phys. Rev. D* **67**, 083507 (2003).
5. C.R. Gould, E.I. Sharapov, S.K. Lamoreaux *Phys. Rev. C* **74**, 024607 (2006). Yu.V. Petrov *et al. Phys. Rev. C* **74**, 064610 (2006); Y. Fujii *et al. Nucl. Phys. B* **573**, 377 (2000).
6. V.V. Flambaum, E.V. Shuryak, physics/0701220.
7. N. Ashby *et al.*, *Phys. Rev. Lett.* **98**, 070802, (2007).
8. W.J. Marciano, *Phys. Rev. Lett.*, **52**, 489 (1984); X. Calmet and H. Fritzsch, *Eur. Phys. J* C24, 639 (2002); P. Langacker, G. Segre and M.J. Strassler, *Phys. Lett. B* **528**, 121 (2002). T. dent, M. Fairbairn. *Nucl. Phys. B* **653**, 256 (2003).
9. J-P. Uzan, *Rev. Mod. Phys.* **75**, **403** (2003).
10. T. Damour and K. Nordtvedt, Phys.Rev.Lett. **70**, 2217 (1993) Phys.Rev.D **48**, 3436 (1993)
11. T. Damour and A. M. Polyakov, *Nucl. Phys. B* **423**, 532 (1994). [arXiv:hep-th/9401069].
12. J. D. Bekenstein, *Phys. Rev. D* **25**, 1527 (1982).
13. H. Sandwick, J. D. Barrow and J. Magueijo, *Phys. Rev. Lett.* **88**, 03102 (2002)
14. K. Olive and M. Pospelov, *Phys. Rev. D* **65**, 085044 (2002)
15. M. T. Murphy, J. K. Webb, V. V. Flambaum. *Mon. Not. R. Astron. Soc.* **345**, 609-638 (2003). J.K. Webb, M.T. Murphy, V.V. Flambaum, V.A. Dzuba, J.D. Barrow, C.W. Churchill, J.X. Prochaska, and A.M. Wolfe, *Phys. Rev. Lett.* **87**, 091301 -1-4 (2001); J. K. Webb, V.V. Flambaum, C.W. Churchill, M.J. Drinkwater, and J.D. Barrow, *Phys. Rev. Lett.*,**82**, 884-887, 1999.
16. V.F. Dmitriev, V.V. Flambaum, J.K. Webb, *Phys. Rev. D* **69**, 063506 (2004).
17. S K Lamoreaux and J R Torgerson *Phys. Rev. D* **69**, 121701 (2004).
18. T. Dent, S. Stern, C. Wetterich, *arXiv:0705.0696.*
19. A. Ivanchik, P. Petitjean, D. Aracil, R. Strianand, H. Chand, C. Ledoux, P. Boisse *Astron. Astrophys.* **440**, 45 (2005). E. Reinhold, R. Buning, U. Hollenstein, A. Ivanchik, P. Petitjean, and W. Ubachs. *Phys. Rev. Lett.* **96**, 151101 (2006).
20. S. Karshenboim, V.V Flambaum, E. Peik, "Atomic clocks and constraints on variation of fundamental constants," in *Springer Handbook of Atomic, Molecular and Optical Physics*, edited by G.W.F. Drake, Springer, Berlin, 2005, Ch. 30, pp455-463; arXiv:physics/0410074.
21. J.D. Prestage, R.L. Tjoelker, and L. Maleki, *Phys. Rev. Lett.* **74**, 3511 (1995).
22. H. Marion *et al.*, *Phys. Rev. Lett.* **90**, 150801 (2003).
23. S. Bize *et al.*, *arXiv:physics/0502117.*
24. E. Peik, B. Lipphardt, H. Schnatz, T. Schneider, Chr. Tamm, S.G. Karshenboim. *Phys. Rev. Lett.* **93**, 170801 (2004).
25. S. Bize *et al.*, *Phys. Rev. Lett.* **90**, 150802 (2003).
26. M. Fischer *et al. Phys. Rev. Lett.* **92**, 230802 (2004).
27. E. Peik, B. Lipphardt, H. Schnatz, T. Schneider, Chr. Tamm, S.G. Karshenboim, *arXiv:physics/0504101.*
28. E. Peik, B. Lipphardt, H. Schnatz, Chr. Tamm, S. Weyers, R. Wynands, *arXiv:physics/0611088.*
29. V.A. Dzuba, V.V. Flambaum, J.K. Webb. *Phys. Rev. A* **59**, 230 (1999).

30. V. A. Dzuba, V. V. Flambaum, M. V. Marchenko, *Phys. Rev. A* **68**, 022506 1-5 (2003).
31. V. A. Dzuba and V. V. Flambaum, *Phys. Rev. A* **72**, 052514 (2005); E. J. Angstmann, V. A. Dzuba, V. V. Flambaum, S. G. Karshenboim, A. Yu. Nevsky, *J. Phys. B* **39**, 1937-1944 (2006) ; physics/0511180,
32. A. T. Nguyen, D. Budker, S. K. Lamoreaux and J. R. Torgerson *Phys. Rev. A.* **69**, 022105 (2004).
33. A. Cingöz et al. *Phys. Rev. Lett.* **98**, 040801, (2007).
34. D.DeMille, invited talk at 35th Meeting of the Division of Atomic, Molecular and Optical Physics, May 25-29, 2004, Tucson, Arizona.
35. V.V. Flambaum, *Phys. Rev. A* **73**, 034101 (2006).
36. van Veldhoven et al. *Eur. Phys. J. D* **31** ,337 (2004).
37. V. V. Flambaum, M. G. Kozlov, arXiv:0705.0849 physics.atom-ph, submitted to Phys. Rev. Lett.
38. V.V. Flambaum, *Phys. Rev. Lett.* **97**, 092502 (2006).
39. E. Peik, Chr. Tamm. *Europhys. Lett.* **61**, 181 (2003).
40. V. A. Dzuba, V.V. Flambaum, J. K. Webb, *Phys. Rev. Lett.* **82**, 888-891, 1999.
41. V.A. Dzuba, V.V. Flambaum, *Phys. Rev. A* **71**, 052509 (2005); V.A. Dzuba, V.V. Flambaum, M.G. Kozlov, and M. Marchenko. *Phys. Rev. A* **66**, 022501-1-8 (2002); J.C. Berengut, V.A. Dzuba, V.V. Flambaum, and M.V. Marchenko. *Phys. Rev. A* **70**, 064101 (2004).
42. R. Srianand, H. Chand, P. Petitjean, and B. Aracil *Phys. Rev. Lett.* 92, 121302 (2004).
43. S. A. Levshakov *et al. Astron.Astrophys***434**, 827 (2005); S. A. Levshakov *et al. Astron.Astrophys***449**, 879 (2006).
44. M.T.Murphy, J.K. Webb, V.V. Flambaum, astro-ph/0612407.
45. J.C. Berengut, V.A. Dzuba, V.V. Flambaum, M.G. Kozlov, M.V. Marchenko, M.T. Murphy, J.K. Webb, *physics/0408017*.
46. M. G. Kozlov, V. A. Korol, J. C. Berengut, V. A. Dzuba, V. V. Flambaum, *Phys. Rev. A* **70**, 062108 (2004); J.C. Berengut, V.A.Dzuba, V.V. Flambaum, M.G. Kozlov. *Phys. Rev. A* **69**, 044102 (2004). J.C. Berengut, V.A. Dzuba, V.V. Flambaum. *Phys. Rev. A* **68**, 022502 1-6 (2003). J. C. Berengut, V. V. Flambaum, and M. G. Kozlov, *Phys. Rev. A* **72**, 044501 (2005); J. C. Berengut, V. V. Flambaum, M. G. Kozlov, *Phys. Rev. A* 73, 012504 (2006).
47. V.A. Dzuba, V.V. Flambaum, *Phys. Rev. A* **61**, 034502 1-3 (2000); E.J. Angstmann, V.A. Dzuba, V.V. Flambaum, *physics/0407141*; E.J. Angstmann, V.V. Flambaum, S.G. Karshenboim. *Phys. Rev. A* **70**, 044104 (2004); E. J. Angstmann, V. A. Dzuba, and V. V. Flambaum *Phys. Rev. A* **70**, 014102 1-4, (2004).
48. A.I. Shlyakhter, *Nature (London)* **264**, 340 (1976); T. Damour, F.J. Dyson *Nucl. Phys. B* **480**, 37 (1996).
49. P. Tzanavaris, J.K. Webb, M.T. Murphy, V.V. Flambaum, and S.J. Curran *Phys. Rev. Lett.* **95**, 041301, 2005.
50. S.G. Karshenboim, *Can. J. Phys.* **78**, 639 (2000).
51. R. G. Helmer and C. W. Reich. *Phys. Rev. C* **49**, 1845 (1994).
52. Z.O. Guimaraes-Filho, O. Helene *Phys. Rev. C* **71**, 044303 (2005).
53. B.R. Beck *et al. Phys. Rev. Lett.* **98**, 142501 (2007).
54. E. V. Tkalya, A. N. Zherikhin, V. I. Zhudov, *Phys. Rev. C* **61**, 064308 (2000); A. M. Dykhne, E. V. Tkalya, *Pis'ma Zh.Eks.Teor.Fiz.* **67**, 233 (1998) [*JETP Lett.* **67**, 251 (1998)].
55. I. D. Moore, I. Ahmad, K. Bailey, D. L. Bowers, Z.-T. Lu, T. P. O'Connor, Z. Yin. Argonne National Laboratory report *PHY-10990-ME-2004*. Y. Kasamatsu, H. Kikunaga, K. Takamiya, T. Mitsugashira, T. Nakanishi, Y. Ohkubo, T. Ohtsuki, W. Sato

and A. Shinohara, *Radiochimica Acta* **93**, 511-514 (2005); S.B. Utter, P. Beiersdorfer, A. Barnes, R. W. Lougheed, J. R. Crespo Lopez-Urrutia, J. A. Becker and M. S. Weiss, *Phys. Rev. Lett.* **82**, 505 (1999).
56. Z.-T. Lu, private communication. E. Peik, private communication.
57. Cheng Chin, V.V. Flambaum, *Phys. Rev. Lett.* **96**, 230801 1-4 (2006).
58. L.Maleki and J.Prestage,*Lecture Notes in Physics*, 648, 341 (2004)
59. S.Schiller et al, *gr-qc/0608081.*
60. A.Bauch and S.Weyers,*Phys.Rev.D65*, 081101R (2002)

MIRROR DARK MATTER

R. FOOT**

School of Physics, University of Melbourne, Victoria 3010 Australia
*** E-mail: rfoot@unimelb.edu.au*

A mirror sector of particles and forces provides a simple explanation of the inferred dark matter of the Universe. The status of this theory is reviewed - with emphasis on how the theory explains the impressive DAMA/NaI annual modulation signal, whilst also being consistent with the null results of the other direct detection experiments.

There is strong evidence for non-baryonic dark matter in the Universe from observations of flat rotation curves in spiral galaxies, from precision measurements of the CMB and from the DAMA/NaI annual modulation signal. The standard model of particle physics has no candidate particles. Therefore new particle physics is suggested.

There are four most basic requirements for a dark matter candidate:

- Massive - The elementary particle(s) comprising the non-baryonic dark matter need to have mass.
- Dark - The dark matter particles couple very weakly to ordinary photons (e.g. electrically neutral particles).
- Stable - The lifetime should be greater than about 10 billion years.
- Abundance - $\Omega_{dark} \approx 5\Omega_b$ (inferred from WMAP CMB observations[1]).

It is not so easy to get suitable candidates from particle physics satisfying these four basic requirements. A popular solution is to hypothesize new neutral particles which are weakly interacting (WIMPs), but this doesn't necessarily make them stable. In fact, the most natural life-time of a hypothetically weakly interacting particle is very short:

$$\tau(wimp) \sim \frac{M_W^4}{g^4 M_{wimp}^5} \sim 10^{-24} \text{ seconds } - \text{ if } M_{wimp} \sim M_Z \ . \quad (1)$$

This is about 41 orders of magnitude too short lived! Of course there is a trivial solution - which is to invent a symmetry to kinematically forbid the particle to decay, but this is ugly because it is ad hoc. The proton and electron, for example, are not stabalized by any such ad hoc symmetry[a]. It is reasonable to suppose that the dark matter particles, like the proton and electron, will also have a good reason for their stability. On the other hand, we also know that the standard model works very well. There is no evidence for anything new (except for neutrino masses). For example, precision electroweak tests are all nicely consistent with no new physics.

A simple way to introduce dark matter candidates which are naturally dark, stable, massive and don't modify standard model physics is to introduce a mirror sector of particles and forces.[2] For every standard model particle there exists a mirror partner[b], which we shall denote with a prime ($'$). The interactions of the mirror particles have the same form as the standard particles, so that the Lagrangian is essentially doubled:

$$\mathcal{L} = \mathcal{L}_{SM}(e, d, u, \gamma, ...) + \mathcal{L}_{SM}(e', d', u', \gamma', ...) \tag{2}$$

At this stage, the two sectors are essentially decoupled from each other except via gravity (although we will discuss the possible ways in which the two sectors can interact with each other in a moment). In such a theory, the mirror baryons are naturally dark, stable and massive and are therefore, a priori, excellent candidates for dark matter. The theory exhibits a gauge symmetry which is $G_{SM} \otimes G_{SM}$ (where $G_{SM} = SU(3)_c \otimes SU(2)_L \otimes U(1)_Y$ is the standard model gauge symmetry).

One can define a discrete symmetry interchanging ordinary and mirror particles, which can be interpreted as space-time parity symmetry ($x \to -x$) if the roles of left and right chiral fields are interchanged in the mirror sector. Because of this geometical interpretation, one cannot regard this discrete symmetry as ad hoc in any sense.

An obvious question is: can ordinary and mirror particles interact with each other non-gravitationally? The answer is YES - but only two terms are consistent with renormalizability and symmetry:[2]

$$\mathcal{L}_{mix} = \frac{\epsilon}{2} F^{\mu\nu} F'_{\mu\nu} + \lambda \phi^\dagger \phi \phi'^\dagger \phi' \ , \tag{3}$$

where $F_{\mu\nu}$ ($F'_{\mu\nu}$) is the ordinary (mirror) $U(1)$ gauge boson field strength tensor and ϕ (ϕ') is the electroweak Higgs (mirror Higgs) field. These two terms are very important, because they lead to ways to experimentally test the idea.

With the above Higgs - mirror Higgs quartic coupling term included, the full Higgs potential of the model has three parameters. Minimizing this potential, one finds that there are two possible vacuum solutions (with each solution holding for a range of parameters): $\langle\phi\rangle = \langle\phi'\rangle \simeq 174$ GeV (unbroken mirror symmetry) and $\langle\phi\rangle \simeq 174$ GeV, $\langle\phi'\rangle = 0$ (spontaneously broken mirror symmetry[c]). While both

[a] Protons and electrons are stabalized by baryon and lepton number $U(1)$ global symmetries which are not imposed, but are accidental symmetries of the standard model. These symmetries cannot be broken by any renormalizable term consistent with the gauge symmetries in the standard model.

[b] For a more comprehensive review, see. e.g. Ref. 3.

[c] Mirror QCD effects eventually break $SU(2) \times U(1)$ in the mirror sector leading to a small, but non-zero VEV for ϕ' in the spontaneously broken case. See Ref. 4 for details.

vacuum solutions are phenomenologically viable, we shall henceforth assume that the mirror symmetry is unbroken, because that case seems more interesting from a dark matter perspective. In the unbroken mirror symmetry case the mass and interactions of the mirror particles are exactly the same as the ordinary particles (except for the interchange of left and right).

Is mirror matter too much like ordinary matter to account for the non-baryonic dark matter in the Universe? After all, ordinary and dark matter have some different properties:

- Dark matter is (roughly) spherically distributed in spiral galaxies, which is in sharp contrast to ordinary matter which has collapsed onto the disk.
- $\Omega_{dark} \neq \Omega_b$ but $\Omega_{dark} \approx 5\Omega_b$.
- Big Bang Nucleosynthesis (BBN) works very well without any extra energy density from a mirror sector.
- Large scale structure formation should begin prior to ordinary photon decoupling.

Clearly there is no 'macroscopic' symmetry. But this doesn't preclude the possibility of exactly symmetric microscopic physics. Why? Because the initial conditions in the Universe might be different in the two sectors. In particular, if in the early Universe, the temperature of the mirror particles (T') were significantly less than the ordinary particles (T) then:

- Ordinary BBN is not significantly modified provided $T' \lesssim 0.5T$.
- $\Omega_{dark} \neq \Omega_b$ since baryogenesis mechanisms typically depend on temperature[d].
- Structure formation in the mirror sector can start before ordinary photon decoupling because mirror photon decoupling occurs earlier if $T' < T$.[7] Detailed studies[8] find that for $T' \lesssim 0.2T$ successful large scale structure follows. This dark matter candidate is also nicely consistent with CMB measurements.[9]
- Furthermore, BBN in the mirror sector is quite different since mirror BBN occurs earlier if $T' < T$. In fact, because of the larger expansion rate at earlier times we would expect that the He'/H' ratio be much larger than the ratio of He/H in the Universe. This would change the way mirror matter evolves on short scales c.f. ordinary matter. Maybe this can explain why mirror matter hasn't yet collapsed onto the disk.[10]

Ok, so mirror matter can plausibly explain the non-baryonic dark matter inferred to exist in the Universe. Can it really be experimentally tested though?

The Higgs mixing term will impact on the properties of the standard model Higgs.[11,12] This may be tested if a scalar is found in experiments, e.g. at the forthcoming LHC experiment. More interesting, at the moment, is the $\epsilon F^{\mu\nu} F'_{\mu\nu}$ term.

[d] The fact that $\Omega_{dark} \neq \Omega_b$ but $\Omega_{dark} \sim \Omega_b$ is suggestive of some similarity between the ordinary and dark matter particle properties, which might be explained within the mirror dark matter context by having exactly symmetric microscopic physics and asymmetric temperatures. For some specific models in this direction, see Ref. 5, 6.

This interaction leads to kinetic mixing of the ordinary photon with the mirror photon, which in turn leads to orthopositronium - mirror orthopositronium oscillations[13] (see also[14]). Null results of current experiments imply[15] $\epsilon < 5 \times 10^{-7}$. Another consequence of the $\epsilon F^{\mu\nu} F'_{\mu\nu}$ term is that it will lead to elastic (Rutherford) scattering of mirror baryons off ordinary baryons, since the mirror proton effectively couples to ordinary photons with electric charge ϵe. This means that conventional dark matter detection experiments currently searching for WIMPs can also search for mirror dark matter![16] The DAMA/NaI experiment already claims direct detection of dark matter.[17] Can mirror dark matter explain that experiment?

The interaction rate in an experiment such as DAMA/NaI has the general form:

$$\frac{dR}{dE_R} = \sum_{A'} N_T n_{A'} \int_{v'_{min}(E_R)}^{\infty} \frac{d\sigma}{dE_R} \frac{f(v', v_E)}{k} |v'| d^3v' \tag{4}$$

where N_T is the number of target atoms per kg of detector, $n_{A'}$ is the galactic halo number density of dark matter particles labeled as A'. We include a sum allowing for more than one type of dark matter particle. In the above equation $f(v', v_E)/k$ is the velocity distribution of the dark matter particles, A', and v_E is the Earth's velocity relative to the galaxy. Also, $v'_{min}(E_R)$ is the minimum velocity for which a dark matter particle of mass $M_{A'}$ impacting on a target atom of mass M_A can produce a recoil of energy E_R for the target atom. This minimum velocity satisfies the kinematic relation:

$$v'_{min}(E_R) = \sqrt{\frac{(M_A + M_{A'})^2 E_R}{2 M_A M_{A'}^2}} \tag{5}$$

The DAMA experiment eliminates the background by using the annual modulation signature. The idea[18] is very simple. The rate, Eq. 4, must vary periodically since it depends on the Earth's velocity, v_E, which modulates due to the Earth's motion around the Sun. That is,

$$R(v_E) = R(v_\odot) + \left(\frac{\partial R}{\partial v_E}\right)_{v_\odot} \Delta v_E \cos\omega(t - t_0) \tag{6}$$

where $\Delta v_E \simeq 15$ km/s, $\omega \equiv 2\pi/T$ ($T = 1$ year) and $t_0 = 152.5$ days (from astronomical data). The phase and period are both predicted! This gives a strong systematic check on their results. Such an annual modulation was found[17] at the 6.3σ Confidence level, with T, t_0 measured to be:

$$\begin{aligned} T &= 1.00 \pm 0.01 \text{ year} \\ t_0 &= 140 \pm 22 \text{ days} \end{aligned} \tag{7}$$

Clearly, both the period and phase are consistent with the theoretical expectations of halo dark matter.

The signal occurs in a definite low energy range from 6 keVee down to the experimental threshold of 2 keVee[e]. No annual modulation was found for $E_R > 6$ keVee. Given that the mean velocity of halo dark matter particles relative to the Earth is of order the local rotational velocity ($\sim$ 300 km/s), this suggests a mass for the (cold) dark matter particles roughly of order 20 GeV, since:

$$E = \frac{1}{2}mv^2 \simeq \frac{m}{20 \text{ GeV}} \left(\frac{v}{300 \text{ km/s}}\right)^2 10 \text{ keV}. \tag{8}$$

Dark matter particles with mass larger than about 60 GeV would give a signal above the 6 keVee region (no such signal was observed in the DAMA experiment). On the other hand, dark matter particles with mass less than about 5 GeV do not have enough energy to produce a signal in the 4-6 keVee energy region - which would be contrary to the DAMA results. Importantly, the mass region sensitive to the DAMA experiment coincides with that predicted by mirror dark matter, since mirror dark matter predictes a spectrum of dark matter elements ranging in mass from hydrogen to iron. That is, with mass GeV $\lesssim M_{A'} \lesssim 55$ GeV. A detailed analysis[16] confirms that mirror dark matter can fit the DAMA experimental data and the required value for ϵ is $\epsilon \sim 10^{-9}$. This fit to the annual modulation signal is given in Fig. 1.

Interestingly, a mirror sector interacting with the ordinary particles with $\epsilon \sim 10^{-9}$ has many other interesting applications (see e.g. Ref. 19, 20). It also consistent with the Laboratory (orthopositronium) bound as well as BBN constraints.[21]

What about the null results of the other direct detection experiments, such as the CDMS, Zeplin, Edelweiss experiments? For any model which explains the DAMA/NaI annual modulation signal, the corresponding rate for the other direct detection experiments can be predicted. These null results do seem to disfavour the WIMP interpretation of the DAMA experiment. However it turns out that they do not, at present, disfavour the mirror dark matter interpretation. Why? because these other experiments are typically all higher threshold experiments with heavier target elements than Na (which, in the mirror matter interpretation, dominates the DAMA/NaI signal) and mirror dark matter has three key features which make it less sensitive (than WIMPs) to higher threshold experiments.

- Mirror dark matter is relatively light $M_H \leq M_{A'} \leq M_{Fe}$.
- The Rutherford cross section has the form:

$$\frac{d\sigma}{dE_R} \propto \frac{1}{E_R^2}$$

  while for WIMPs it is E_R independent (excepting the energy dependence of the form factors).

[e]The unit, keVee is the so-called electron equivalent energy, which is the energy of an event if it were due to an electron recoil. The actual nuclear recoil energy (in keV) is given by: keVee/q, where q is the quenching factor ($q_I \simeq 0.09$ and $q_{Na} \simeq 0.30$).

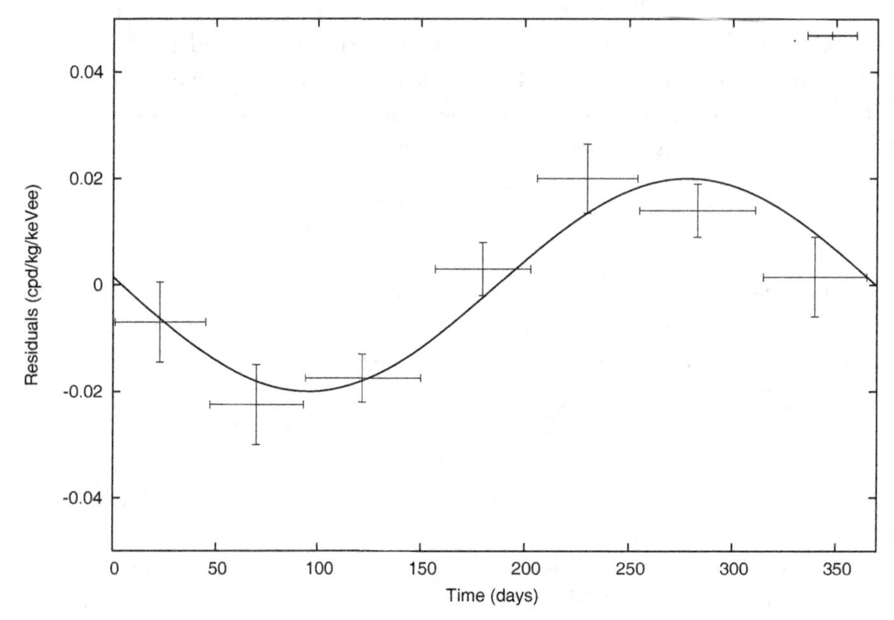

Fig. 1. DAMA/NaI annual modulation signal (taking data from Ref. 17) together with the mirror matter prediction. Note that the initial time in this figure is August 7th.

- Mirror particles interact with each other. This implies that the Halo particles are in local thermodynamic equilibrium, so that e.g. $T = \frac{1}{2}M_{H'}\overline{v_{H'}^2} = \frac{1}{2}M_{O'}\overline{v_{O'}^2}$ ($\approx$ 300 eV assuming the standard assumptions of an isothermal halo in hydrostatic equilibrium[3]). Thus heavier elements have smaller mean velocities.

To summarize, having a mirror sector is a simple way to explain the inferred dark matter of the Universe. There is experimental support for this particular dark matter hypothesis, coming from the positive DAMA annual modulation signal. We must await future experiments to see if this explanation is the correct hypothesis.

Acknowledgments

This work was supported by the Australian Research Council.

References

1. D. N. Spergel *et al.*, (WMAP Collaboration), *Astrophys. J. Suppl.* **148**, 175 (2003) [astro-ph/0302209].
2. R. Foot, H. Lew and R. R. Volkas, *Phys. Lett.* **B272**, 67 (1991). The idea that a mirror sector might exist was discussed earlier, prior to the advent of the standard model of particle physics, in T. D. Lee and C. N. Yang, *Phys. Rev.* **104**, 256 (1956); I. Kobzarev, L. Okun and I. Pomeranchuk, *Sov. J. Nucl. Phys.* **3**, 837 (1966); M. Pavsic, *Int. J. Theor. Phys.* **9**, 229 (1974). The first application to non-baryonic dark matter was given in: S. I. Blinnikov and M. Yu. Khlopov, *Sov. J. Nucl. Phys.* **36**, 472 (1982); *Sov. Astron.* **27**, 371 (1983).

3. R. Foot, *Int. J. Mod. Phys.* **D13**, 2161 (2004) [astro-ph/0407623].
4. R. Foot and H. Lew, hep-ph/9411390; R. Foot, H. Lew and R. R. Volkas, *JHEP* **0007**, 032 (2000) [hep-ph/0006027].
5. L. Bento and Z. Berezhiani, *Phys. Rev. Lett.* **87**, 231304 (2001) [hep-ph/0107281]; hep-ph/0111116.
6. R. Volkas and R. Foot, *Phys. Rev.* **D68**, 021304 (2003) [hep-ph/0304261]; *Phys. Rev.* **D69**, 123510 (2004) [hep-ph/0402267].
7. Z. Berezhiani, D. Comelli and F. L. Villante, *Phys. Lett.* **B503**, 362 (2001) [hep-ph/0008105].
8. A. Yu. Ignatiev and R. R. Volkas, *Phys. Rev.* **D68**, 023518 (2003) [hep-ph/0304260].
9. Z. Berezhiani *et al.*, *Int. J. Mod. Phys.* **D14**, 107 (2005) [astro-ph/0312605]; P. Ciarcelluti, astro-ph/0312607; P. Ciarcelluti, *Int. J. Mod. Phys.* **D14**, 187 (2005) [astro-ph/0409630]; *Int. J. Mod. Phys.* **D14**, 223 (2005) [asto-ph/0409633].
10. R. Foot and R. R. Volkas, *Phys. Rev.* **D70**, 123508 (2004) [astro-ph/0407522].
11. R. Foot, H. Lew and R. R. Volkas, *Mod. Phys. Lett.* **A7**, 2567 (1992).
12. A. Yu. Ignatiev and R. R. Volkas, *Phys. Lett.* **B487**, 294 (2000) [hep-ph/0005238].
13. S. L. Glashow, *Phys. Lett.* **B167**, 35 (1986).
14. R. Foot and S. N. Gninenko, *Phys. Lett.* **B480**, 171 (2000) [hep-ph/0003278].
15. R. Foot, *Int. J. Mod. Phys.* **A19**, 3807 (2004) [astro-ph/0309330].
16. R. Foot, *Phys. Rev.* **D69**, 036001 (2004) [hep-ph/0308254]; R. Foot, *Mod. Phys. Lett.* **A19**, 1841 (2004) [astro-ph/0405362]; R. Foot, *Phys. Rev.* **D74**, 023514 (2006) [astro-ph/0510705].
17. R. Bernabei *et al.*, (DAMA Collaboration), *Riv. Nuovo Cimento.* **26**, 1 (2003) [astro-ph/0307403]; *Int. J. Mod. Phys.* **D13**, 2127 (2004) and references therein.
18. A. K. Drukier, K. Freese and D. N. Spergel, *Phys. Rev.* **D33**, 3495 (1986); K. Freese, J. A. Frieman and A. Gould, *Phys. Rev.* **D37**, 3388 (1988).
19. R. Foot and S. Mitra, *Astropart. Phys.* **19**, 739 (2003) [astro-ph/0211067].
20. R. Foot and Z. K. Silagadze, *Int. J. Mod. Phys.* **D14**, 143 (2005) [astro-ph/0404515].
21. E. D. Carlson and S. L. Glashow, *Phys. Lett.* **B193**, 168 (1987).

ANALYTIC, NON-PERTURBATIVE, ALMOST-EXACT QED: THE TWO-POINT FUNCTION

H. M. FRIED*

Physics Department, Brown University,
Providence, RI 02912 USA
** E-mail: fried@het.brown.edu*

The goal, of course, is QCD; but one must learn how to walk before one can run. This analysis is in four dimensions, using the Minkowski metric and a modified Fradkin representation of a modified Schwinger functional solution for QED. And this is a work-in-progress presentation of techniques and results discovered within the last year, which so far have treated the sum of all Feynman graphs for the dressed electron and photon propagators in quenched approximation. The initial feature which makes such calculations possible is the use of a special, relativistic, photon gauge, which provides a framework for the scaling transformations of the gauge-dependent electron propagator, and the application of strict current conservation for the photon propagator. In this (rapid) presentation, electron mass-renormalization will be suppressed, although its wave-function renormalization will be clear.

Begin with a modified Schwinger representation[1] for the dressed electron propagator in quenched approximation,

$$S_c'(x-y) \equiv e^{\mathcal{D}_A} \cdot G_c(x,y \mid A) \, \frac{e^{L[A]}}{< S >} \, |_{A\to 0} \to e^{\mathcal{D}_A} \, G_c(x,y|A) \, |_{A\to 0} \, , \tag{1}$$

where

$$\mathcal{D}_A = -\frac{i}{2} \int\!\!\int \frac{\delta}{\delta A_\mu} \, D_c^{\mu\nu} \, \frac{\delta}{\delta A_\nu}, \, [m_1 + \gamma \cdot (\partial_x - igA(x))] \, G_c(x,y|A) = \delta^{(4)}(x-y) \, ,$$

$$\tilde{\mathcal{D}}_c^{\mu\nu}(k) = \frac{1}{k^2 - i\epsilon} \left[\delta_{\mu\nu} - \rho \frac{k_\mu k_\nu}{k^2 - i\epsilon} \right] \, , \; L[A] = \mathrm{Tr}\, \ln\left[1 - ig\gamma \cdot A S_c\right] \, ,$$

$$S_c = G_c[A] \, |_{A\to 0} \, , \; < S > \equiv e^{\mathcal{D}_A} \cdot e^{L[A]} \, |_{A\to 0} \, .$$

Then introduce the exact Fradkin representation[2] for the Green's function $G_c(x,y|A)$ corresponding to the propagator of a relativistic electron in a 4-vector potential field $A_\mu(x)$,

$$G_c(x,y|A) = i \int_0^\infty ds \, e^{-ism_0^2} \cdot e^{i \int_0^s ds' \, \frac{\delta^2}{\delta v_\mu^2(s')}} \cdot \left(m_0 - \gamma \cdot \frac{\delta}{\delta v(s)} \right)$$
$$\cdot \delta^{(4)} \left(x - y + \int_0^s ds' v(s') \right) \cdot$$

$$\cdot e^{-ig\int_0^s ds' v_\mu(s') A_\mu(y-\int_0^{s'} v)} \cdot \left(e^{g\int_0^s ds' \sigma_{\mu\nu} F_{\mu\nu}(g-\int_0^{s'} \nu)} \right)_+ \Big|_{v_\mu \to 0} \cdot \quad (2)$$

The existence of the ordered exponential (OE) in (2) is the main reason why it has never been possible to formulate an analytic, non-perturbative mechanism for solving QED, for analytic approximations to general OEs have only been given in adiabatic and stochastic limits,[3] which are not applicable to the requirements of (2). It will be useful to apply a functional translation operator to the OE of (2) in order to extract it's A-dependence, and so be able to perform the linkage operation of (1). With the convenient variable change $u(s') = \int_0^{s'} ds'' v(s'')$, and with $z = x - y, u(s) + z = 0, u(0) = 0$, and with $S_c^{'}(z) ='' (m - \gamma \cdot \partial)'' \mathcal{S}_c^{'}(z)$, one obtains

$$\begin{aligned} \mathcal{S}_c^{'}(z) = i \int_0^\infty ds\, e^{-ism_0^2} e^{-\frac{1}{2}\mathrm{Tr}\ln(2ah)} \cdot N^{'} \cdot \int d[u] e^{\frac{i}{2}\int\int_0^s u(2ah)^{-1}u} \cdot \delta^{(4)}(z + u(s)) \\ \cdot e^{\frac{ig^2}{2}\int\int_0^s ds_1 ds_2 u_\mu^{'}(s_1) \mathcal{D}_c^{\mu\nu}(u(s_1)-u(s_2)) u_\nu^{'}(s_2)} \\ e^{+2ig^2 \int\int_0^s ds_1 ds_2 \frac{\delta}{\delta X_{\mu\nu}(s_1)} \partial_\mu \partial_{\mu'} \mathcal{D}_c^{\nu\nu'}(\zeta) \frac{\delta}{\delta X_{\mu'\nu'}(s_2)}} \\ e^{+g^2\int\int_0^s ds_1 ds_2 u_\mu^{'}(s_1) \vec{\partial}_\nu \mathcal{D}_c^{\mu\lambda}(\zeta) \frac{\delta}{\delta X_{\nu\lambda}(s_2)}} \cdot \left(e^{\int_0^s ds' \sigma_{\mu\nu} X_{\mu\nu}(s^1)} \right)_+ \Big|_{X\to 0} \end{aligned} \quad (3)$$

where $\zeta_\mu \to u_\mu(j_1) - u_\mu(j_2) \equiv \Delta u(s_1, s_2), h(s_1, s_2) = \frac{1}{2}(s_1 + s_2 - (|s_1 - s_2|))$, N' is a normalization constant depending on the Δs partitions of the functional integral, and a is a real, positive number to be set equal to 1 at the end of the calculation. The notation "$(m - \gamma \cdot \partial)$" means that a mass-renormalization δm has been defined, but suppressed for this presentation, since the object of true interest is the remaining $S_c'(z)$.

One may now make a first, and somewhat remarkable observation: Because of the nature of the Dirac $\sigma_{\mu\nu} = \frac{1}{4}[\gamma_\mu, \gamma_\nu]$, and the asymmetry of the$X_{\mu\nu}$, one can prove that, to all g^2 orders,

$$e^{+2ig^2\int\int_0^s \frac{\delta}{\delta X}(\partial\partial D_c)\frac{\delta}{\delta X}} \cdot \left(e^{\int_0^s \sigma X} \right)_+ \Big|_{X\to 0} = 1\,. \quad (4)$$

Equation (4) was a surprise, first seen in simple demonstrations that its terms of order g^2, g^4, g^6 all vanished by algebraic cancellation; and with a proof that this is true for arbitrary order now in hand. However, the $u' \cdots (OE)$ linkages do not appear to vanish in a similar way, but generate log divergent terms in every order; and the question of how to handle these in a non-perturbative way remains.

A second and most useful observation may now be made: A general relativistic gauge may be defined by a parameter of the (bare) photon propagator in momentum space,

$$\tilde{\mathcal{D}}_c^{\mu\nu}(k) = \frac{1}{k^2 - i\epsilon}\left(\delta_{\mu\nu} - \rho \frac{k_\mu k_v}{k^i - i\epsilon} \right)$$

with $\rho = 0, 1$, and -2 defining the Feynman, Landau, and Yennie gauges, respectively. In configuration space, this becomes

$$\mathcal{D}_c^{\mu\nu}(z) = \frac{i}{4\pi^2}\left(\frac{\delta_{\mu\nu}[1-\rho/2]}{z^2+i\epsilon} + \rho\frac{z_\mu Z_\nu}{(z^2+i\epsilon)^2}\right).$$

A most useful, special gauge is here defined by the choice: $\rho = +2 : \mathcal{D}_c^{\mu\nu}(z) = \frac{i}{2\pi^2}\frac{z_\mu z_\nu}{(z^2 i\epsilon)^2}$.

Why is this gauge useful? Because the $u' \cdots u'$ term of (3) may then be rewritten in the form

$$\exp\left[-\frac{g^2}{4\pi^2}\int\int_0^s ds_1 ds_2 \frac{(u_\mu'(u)\cdot\Delta u_\mu)}{[(\Delta u)^2+i\epsilon]}\cdot\frac{(\Delta u_v \cdot u_v'(s_2))}{[(\Delta u)^2+i\epsilon]}\right]$$
$$= \exp\left[\gamma\int\int_0^s ds_1 ds_2 \frac{\partial}{\partial s_1}\ln Z\cdot\frac{\partial}{\partial s_2}\ln Z\right], \quad (5)$$

with $Z = M^2\left[(\Delta u)^2 + i\epsilon\right]$, $\gamma = \frac{g^2}{16\pi^2}$, and M an arbitrary (for the moment) mass parameter introduced for dimensional reasons. This is almost a perfect (double) differential, but not quite; we can, however, rewrite it as

$$\exp\left[\gamma\int\int_0^s ds_1 ds_2\left\{\frac{1}{2}\frac{\partial}{\partial s_1}\frac{\partial}{\partial s_2}\ln^2 Z - \ln Z\cdot\frac{\partial}{\partial s_1}\frac{\partial}{\partial s_2}\ln Z\right\}\right], \quad (6)$$

in which the first term of (6) is a perfect differential, and as such the u-fluctuations of the functional integral all cancel away, and the result is given by the end-point $u(s) = -z, u(0) = 0$ quantities as

$$= \exp\left[-\gamma\ln^2\left(\frac{z^2+i\epsilon}{i\epsilon}\right) - 2\gamma\ln\left(\frac{z^2+i\epsilon}{i\epsilon}\right)\cdot\ln(i\epsilon M^2)\right]. \quad (7)$$

Evaluation of the second term of (6) was first attempted by approximation: Since the ϵ of $\ln[Z]$ acts as a cut-off parameter in configuration space (in momentum space, $\epsilon \sim \Lambda^{-2}$), one expects that $\ln[Z]$ should be "slowly varying", and the second term of (6) could reasonably be approximated by

$$\exp\left[-\gamma\langle\ln Z\rangle\int\int_0^s ds_1 ds_2\frac{\partial^2}{\partial s_1\partial s_2}\ln Z\right], \quad (8)$$

where the ¡average¿is taken over $s_{1,2}$ and over the u-fluctuations. The integrals of (8) are now perfect differentials, which can be evaluated immediately,

$$\int\int_0^s ds_1 ds_2\frac{\partial^2}{\partial s_1\partial s_2}\ln Z = -2\ln\left(\frac{z^2+i\epsilon}{i\epsilon}\right), \quad (9)$$

and the remaining functional integral is trivial, yielding the free-particle result:

$$I_0 = (4\pi a s)^{-2}\exp[iz^2/4as].$$

This procedure was originally thought to be the beginning of a strong-coupling approximation; but one can do much better, as follows.

Add and subtract to the second term of (6) the quantity

$$-\gamma \ln Q \cdot \int \int_0^s ds_1 ds_2 \frac{\partial^2}{\partial s_1 \partial s_2} \ln Z \,, \tag{10}$$

where Q is a real, positive number > 1; in the remaining functional integral

$$\begin{aligned}\mathcal{R}(z^2, a) &= e^{-\frac{1}{2}\mathrm{Tr}\ln(2ah)} \cdot N' \int d[u] e^{\frac{i}{2}\int\int u(2ah)^{-1}u} \cdot \delta^{(4)}(z + u(s)) \\ &\cdot e^{-\gamma \int\int \ln Z \cdot \frac{\partial^2}{\partial s_1 \partial s_2} \ln Z} \cdot e^{-g^2 \int\int ds_1 ds_2 u'(s_1) \vec{\partial}_\nu (\mathcal{D}^{\mu\lambda} \cdot) \frac{\delta}{\delta X_{\nu\lambda}}} \left(e^{\int \sigma \cdot x}\right)_+ \Big|_{X \to o} \,,\end{aligned} \tag{11}$$

this replaces the term $e^{-\gamma \int\int \ln Z \cdot \frac{\partial^2}{\partial s_1 \partial s_2} \ln Z}$ by

$$e^{-\gamma \ln Q \cdot \int\int ds_1 ds_2 \frac{\partial^2}{\partial s_1 \cdot \partial s_2} \ln Z} \cdot e^{-\gamma \int\int ds_1 ds_2 \ln(Z/Q) \frac{\partial^2}{\partial s 1 \cdot \partial s_2} \ln(Z/Q)} \,. \tag{12}$$

But $\ln[Z/Q]$ can be written as $\ln\left(M^2\left[\frac{(\Delta\mu)^2 + i\epsilon}{Q}\right]\right) \Rightarrow \ln\left(M^2\left[\frac{(\Delta\mu)^2}{Q} + i\epsilon\right]\right)$, and rescaling the dummy variables $u_\mu(s_i) \to \sqrt{Q}\bar{u}_\mu(s_i)$ consistently in (11) produces the scaling statement

$$\mathcal{R}(z^2, a) = Q^{-2} \cdot e^{2\gamma \ln Q \cdot \ln\left(\frac{z^2 + i\epsilon}{i\epsilon}\right)} \cdot I\left(\frac{z^2}{Q}, \frac{a}{Q}; \frac{1}{Q}\right) \tag{13}$$

where

$$\begin{aligned}I\left(\frac{z^2}{Q}, \frac{a}{Q}, \frac{1}{Q}\right) &= e^{-\frac{1}{2}\mathrm{Tr}\ln(z \frac{a}{Q} h)} \cdot N' \int d(\bar{u}) e^{\frac{i}{2}\int\int \bar{u}(2\frac{ah}{Q})^{-1}\bar{u}} \cdot \delta^{(4)}\left(\bar{u}(s) + \frac{z}{\sqrt{Q}}\right) \\ &\cdot e^{-\gamma \int\int \ln Z(\Delta\bar{u}) \cdot \frac{\partial^2}{\partial s_1 \partial s_2} \ln Z(\Delta\bar{u})} \\ &\cdot e^{+\frac{g^2}{Q} \cdot \int\int \bar{u}'_\mu(s_1) \vec{\partial}_\nu \mathcal{D}_c^{\mu\lambda}(\Delta\bar{u}) \frac{\delta}{\delta X_{\mu\lambda}(s_2)}} \left(e^{\int \sigma \cdot X}\right)_+ |_{X \to 0} \,.\end{aligned} \tag{14}$$

Because of the $1/Q$ dependence of the OE term, this is not a useful scaling relation. But if Q is taken as arbitrarily large- far larger than any cut off associated with the logarithmically divergent $u' \cdots (OE)$ terms - then all of those term are individually removed, and their sum appears in the exponential factor: $\exp\left[2\gamma \ln Q \cdot \ln\left(\frac{z^2 + i\epsilon}{i\epsilon}\right)\right]$. This represents a magnificent calculational tool, for now the OE dependence, which has always blocked non-perturbative estimations (except in Bloch-Nordsieck approximations, where such terms are neglected), has been effectively summed. And now, with $R(z^2, a)$ independent of Q, one has a useful scaling relation:

$$\mathcal{R}(z^2, a) = Q^{-2} \cdot e^{2\gamma \ln Q \cdot \ln\left(\frac{z^2 + i\epsilon}{i\epsilon}\right)} \cdot I\left(\frac{z^2}{Q}, \frac{a}{Q}\right) . \tag{15}$$

It should be noted that the OE terms are themselves gauge-invariant, since they stem from an initial $F_{\mu\nu}$ dependence; but this special gauge provides the framework for their summation.

Renormalization Group methods can now be employed to provide a differential equation (DE) for $I(\frac{z^2}{Q}.\frac{a}{Q})$, by considering small variations of the very large Q, and calculating $0 = Q\frac{\partial}{\partial Q}\mathcal{R}(z^2.a)$,

$$0 = \left[2\gamma \ln\left(\frac{z^2 + i\epsilon}{i\epsilon}\right) - 2\right] I\left(\frac{z^2}{Q}, \frac{a}{Q}\right) - \left(z^2\frac{\partial}{\partial z^2} + a\frac{\partial}{\partial a}\right) I\left(\frac{z^2}{Q}, \frac{a}{Q}\right). \quad (16)$$

It is simplest to rescale $z^2 \to z^2 Q$, $a \to aQ$ to obtain the DE

$$\left[2\gamma \ln Q + 2\gamma \ln\left(\frac{z^2 + i\epsilon}{i\epsilon}\right) - 2\right] I(z^2, a) = \left(z^2\frac{\partial}{\partial z^2} + a\frac{\partial}{\partial a}\right) I\left(z^2, a\right). \quad (17)$$

It is appropriate to look for a solution of form $I(z^2,a) = I_0(z^2,a)\cdot J(z^2,a)$. where $\mathcal{S}'(z)|_{g\to 0} = I_0(z^2,a)$, because out of the "landscape" of possible solutions to the functional Schwinger equations, this is the one desired. Further, one can demand that

$$J(z^2, a)\ |_{a\to 1, z^2 \to 0} \Rightarrow 1\,,$$

so that the equal-time, anti-commutation relations originally assumed for free and dressed fermion field operators are the same. With $J = \exp[\Omega]$(17) becomes

$$\left(z^2\frac{\partial}{\partial z^2} + a\frac{\partial}{\partial a}\right)\Omega(z^2, a) = 2\gamma\left[\ln Q + \ln\left(\frac{z^2 + i\epsilon}{i\epsilon}\right)\right], \quad (18)$$

and this is the relation one must now solve.

Inspection shows that to any solution of (18) satisfying the above "boundary conditions" can be added an almost arbitrary function of z^2/a, which sum will produce another solution to (18). This lack of mathematical uniqueness, however, is not detrimental to our physical description, for there is yet one more condition which must be applied and interpreted; and the final result will have a satisfactory "physical uniqueness". At this point one can realize that, whatever the solution chosen, the essence of this special gauge + rescaling method is that the non-perturbative solutions to this dressed propagator are essentially multiplicative in coordinate space. In contrast, as we all know, Feynman graphs in higher orders of perturbation theory involve horrendous and overlapping integrals in momentum space. This special gauge is the bridge that leads to analytically-obtainable exact solutions in configuration space. Of course, Fourier transforms must finally be taken; but that is a separate matter.

At the end of this calculation, the parameter $a \to 1$, and any a-dependence in the solution becomes a constant (a general constant will be chosen below). Since there is no a-dependence on the RHS of (18), one may take as the simplest solution that obtained by assuming $\Omega \to \Omega(y), y = \ln\left[M^2(z^2 + i\epsilon)\right]$, and (18) then becomes $\frac{d\Omega}{dy} = 2\gamma\left[y + \ln(\frac{Q}{i\epsilon M^2})\right]$ with solution $\Omega(y) = \gamma y^2 + 2\gamma y \ln(\frac{Q}{i\epsilon M^2}) + \kappa, \quad \kappa = \text{const.}$, so that

$$J \to \exp\left[\gamma \ln^2(M^2(z^2 + i\epsilon)) + 2\gamma \ln\left[M^2(z^2 + i\epsilon)\right]\cdot \ln\left(\frac{Q}{i\epsilon M^2}\right) + \kappa\right]. \quad (19)$$

Remembering to rescale $z^2 \to z^2/Q$ (in order to undo the passage from (16) to (17), and including the previous, perfect-differential term of (7),

$$\frac{\gamma}{2}\int\int_0^s ds_1 ds_2 \frac{\partial^2}{\partial s_1 \partial s_2} \ln^2 Z = -\gamma \ln^2\left(\frac{z^2+\epsilon}{i\epsilon}\right) - 2\gamma \ln\left(\frac{z^2+i\epsilon}{i\epsilon}\right) \cdot \ln(i\epsilon M^2)\,,$$

the entire answer becomes

$$\begin{aligned} \mathcal{S}_c'(z) = I_0(z^2,1) \cdot \exp\Big\{ 2\gamma \ln\left(\frac{z^2+i\epsilon}{i\epsilon}\right) \cdot \ln\left(\frac{Q}{i\epsilon M^2}\right) \\ + 2\gamma \ln Q \cdot \ln(i\epsilon M^2) - \gamma \ln^2(i\epsilon\kappa^2) + \kappa \Big\}. \end{aligned}$$

If the exponential factor multiplying I_0 is to become unity when $z^2 \to 0$, then the constant of this solution must be chosen as $\kappa = \gamma \ln^2(i\epsilon M^2) - 2\gamma \ln Q \cdot \ln(i\epsilon M^2) - is\delta m^2$, where the term is δm^2 has been included in this constant (obtained from our previous, suppressed knowledge of δm) so as to renormalize the bare mass sitting in the exponential factor of (2). The entire result, with m_0 everywhere replaced by m, is then

$$\mathcal{S}_c'(z) = I_0(z^2,1) \cdot \exp\left[2\gamma \ln\left(\frac{z^2+i\epsilon}{i\epsilon}\right) \cdot \ln\left(\frac{Q}{i\epsilon M^2}\right)\right], \tag{20}$$

where the result of all interactions is, in configuration space, simply a multiplicative, log divergent, exponential factor multiplying the free-field propagator.

Without actually performing the Fourier transform into momentum space, let us try to guess the electron's wave-function renormalization (WFR) constant. Going to the mass shell in momentum space corresponds to taking the limit $z^2 \to \infty$ in coordinate space. This can be represented as an IR limit in momentum space, if z^2 is represented by $1/\mu^2$; also, as noted above, ϵ can be thought of as an UV cut-off in momentum space, $\epsilon \sim \Lambda^{-2}$. The argument of the multiplicative exponential factor of (20) then becomes

$$2\gamma\left[-\left(\frac{\pi}{2}\right)^2 + \ln\left(\frac{\Lambda^2}{\mu^2}\right)\ln\left(\frac{Q\Lambda^2}{\mu^2}\right)\right] - i\frac{\pi}{2}(2\gamma)\cdot \ln\left(\frac{\Lambda^4 Q}{\mu^2 M^2}\right),$$

and in order for the Z_2 of this solution to be real, one must choose $\left(\Lambda^4 Q/\mu^2 M^2\right) = 1$, or $M^2 = Q\Lambda^4/\mu^2$, so that $Z_2 = \exp\left[-2\gamma\left((\frac{\pi}{2})^2 + \ln^2(\frac{\Lambda^2}{\mu^2})\right)\right]$ is not only real, but is bounded between 0 and 1, as required by the formal theory.

Finally, one can argue that the lack of mathematical uniqueness of the solution (19) is really not physically relevant, because the only function of the dressed electron propagator is to produce a WFR constant which can be identified as needed in every n-point function, so as to cancel with an equal Z_1, and help provide a gauge-independent renormalization of the electron's charge; Z_2 is not a measurable quantity. Any other solution chosen in place of (19), which satisfies the above $g^2 \to 0$

and $z^2 \to 0$ requirements, will just correspond to a change in the unmeasurable Z_2. The really relevant points of this analysis are that (i) it provides a straightforward method of choosing a non-perturbative Z_2 which can be used in conjunction with a similar analysis of higher n-point functions to produce gauge-invariant, non-perturbative results; and (ii) this analysis of the two-point function suggests that one will again find a simple description of non-perturbative Physics in configuration space, in conjunction with the use of the special gauge.

Exactly this second expectation is realized by current work-in-progress on the dressed photon propagator, where the sum over all virtual photons exchanged across a closed electron loop is provided by the gauge-invariant relation

$$g^2 \cdot e^{\mathcal{D}_A} \mathrm{tr}\left[\gamma_\mu G_c(x,y|A)\gamma_\nu G_c(y,x|A)\right]\Big|_{A_\mu \to 0} . \tag{21}$$

When a representation and an analysis identical to that used above is employed for each of the Green's functions of (21), after the linkage operations and rescaling are performed, one finds that all the gauge-dependent pieces of the electron-propagator cancel away, leaving a pair of functional integrals which satisfy a simple scaling relation. However, the inhomogeneous z^2 dependence, analogous to that of the RHS of (18), is missing, which requires some further input mechanism in order to generate a solution. That needed mechanism is the requirement of gauge-invariance (more properly, current conservation) in every perturbative order of the expansion of the functional integrals corresponding to (21). One is able to convert each of those perturbative requirements into a differential equation, such that the sum of all relevant perturbative quantities satisfies the very same form of equation. And that equation can be solved (with one arbitrary constant which can be chosen in a unique manner) and as such provides a non-perturbative, configuration-space solution for (21).

Work is now beginning on the application of these techniques to the vertex function of QED; but before any detailed study is begun, one can foresee certain relevant properties of its non-perturbative representation. It has always seemed somewhat suggestive that eikonal representations of the QED vertex function,[4] as well as the sum of its leading perturbative contributions[5] should produce an exponential of either a *log* or a $[log]^2$ dependence on momentum transfer (depending upon the way in which large virtual momenta are limited); but this has now become clear from the forms found using the special gauge. What will surely happen in this coming calculation of the QED vertex, to all orders of its coupling constant, is that the gauge dependent exponential factors will cancel away, leaving (after Fourier transforms into momentum space are performed) the exponential of finite, logarithmic dependence upon relevant momentum transfer.

Finally, a word on the extension to QCD. Color complications may be expected, compared to the simpler, Abelian QED, but there are both approximations and additional Gaussian integrals which can be employed to bring a measure of success to the enterprise. One must use functional integrals which avoid integration over gauge copies, as well as introduce techniques to handle the ever-present OEs of the theory; and one expects that such complications can be overcome. The non-Abelian gauge groups will require current conservation relations between functional representations of different n-point functions, rather than involving only the same n-point function, as in the QED photon propagator. But as long as the (bare) gluons are massless, a special gauge can be found for their propagator's coordinate space representations; and the main structure of the QED analysis suggested above should be possible. This direct approach to non-perturbative solutions of 4-dimensional QCD may be well worth trying.

References

1. H. M. Fried, "Functional Methods and Models in Quantum Field Theory", The MIT Press, Cambridge, MA (1972): Chapter 3.
2. H. M. Fried, "Green's Functions and Ordered Exponentials", Cambridge University Press, UK (2002); Chapter 3.
3. A. Dykne, *Sov. Physics JETP* 11 (1960); *JETP Lett.* **14**, 941 (1962); L. Aspinall and J. C. Percival, *Proc. R. Soc.* **90**, 315 (1967), and references therein. M. S. Brachet and H. M. Fried, *J. Math. Phys.* **28**, 15 (1987); *Phys. Lett.* **103A**, 309 (1984); H. M. Fried, *J. Math. Phys.* **28**, 1275 (1987); **30**, 1161 (1999).
4. H. M. Fried and T. K. Gaisser, *Phys. Rev.* **1798**, 1491 (1969).
5. R. Jackiw, *Ann. Phys. (N.Y.)* **48**, 292 (1968).

STERILE NEUTRINOS IN A 6×6 MATRIX APPROACH

T. GOLDMAN*

Theoretical Division, Los Alamos National Laboratory,
Los Alamos, New Mexico 87545, USA
**E-mail: tgoldman@lanl.gov*
www.lanl.gov

Quark-lepton symmetry invites consideration of the existence of sterile neutrinos. Long ago, we showed that this approach predicts large neutrino mixing amplitudes. Using a Weyl spinor approach, we show, in an analytic example, how this, and pseudo-Dirac pairing, can develop within a reduced rank version of the conventional see-saw mechanism, from small intrinsic mixing strengths. We show by numerical examples that mixing of active and sterile neutrinos can affect the structure of oscillations relevant to extraction of neutrino mixing parameters from neutrino oscillation data.

Keywords: McKellar; Neutrinos; Sterile; Mass; Festschrift.

1. Quark-Lepton Symmetry is Our Basis

In the Standard Model (SM) as first formulated, there were no neutrino mass terms as no right-chiral projections of Dirac neutrino fields were known to exist. Excluding them left no means to produce Dirac neutrino mass terms and Majorana mass terms required introduction of either non-renormalizable terms in the Lagrangian or new scalar fields with unit weak isospin.

However, the formulation of Grand Unified Theories (GUTs) in the mid-70's, $SU(5)$ in particular,[1] made clear that the fundamental degrees of freedom were not Dirac bispinors but two-component Weyl spinors. Later developments in supersymmetry and supergravity amplified this contention. In the Weyl spinor basis, all known fermions, except the neutrinos, appeared in left- and right-chiral pairs ($(\frac{1}{2}, 0)$ and $(0, \frac{1}{2})$ irreps under the Lorentz group). The pairing, along with equal mass terms, were necessary to allow construction of Dirac bispinors which could satisfy the known (to high accuracy) conservation of electric and color charges. Thus, right-chiral partners for the known neutrinos were not required, but, to some of us, at least, seemed strongly invited, especially as the successes of quark-lepton symmetry grew over the following decade: charm,[2] and then after the discovery[3] of the τ-lepton, the b-quark[4] and eventually the t-quark.[5]

2. See-Saw Mechanism

At Los Alamos, a number of researchers and visitors, including Stephenson, Slansky, Ramond and Gell-Mann,[6] recognized that the right-chiral fields were unconstrained, even in $SU(5)$, in the possible Majorana (or as we prefer to say here, Weyl) mass term possible – the mass could be as large as the GUT scale ($M \sim 10^{16}$ GeV). Furthermore, this, combined with now normal (order quark or charged lepton) Dirac mass terms ($m \sim 10^{\pm 3}$ GeV) that should appear, would produce eigenstates with very small Majorana masses ($\sim m^2/M$) and that were almost purely left-chiral neutrinos, that is, those that participate in the weak interactions. As a bonus, the known Cabibbo-Kobayashi-Maskawa (CKM) mixing[7] between quark mass and weak interaction eigenstates strongly suggested that similar physics should develop in the lepton sector, producing the long-conjectured oscillation of neutrinos[8] (although between different flavors rather than between particle and antiparticle as originally suggested).

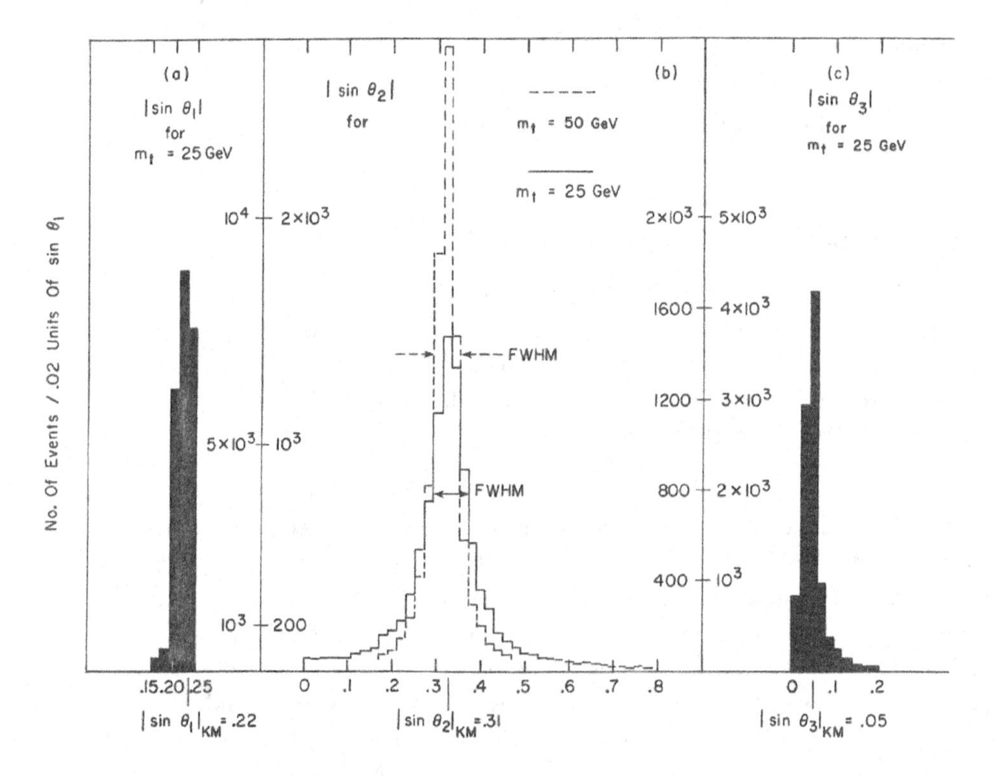

Fig. 1. Mixing angle distributions for random mass matrix entries from Ref. 9.

2.1. *An early effort*

In the absence of any credible detailed conjectures as to the structure of the mass matrices in the lepton sector (although there were a plethora of papers about what would now be called "textures"), we[9] carried out a Monte Carlo study (popular more recently in considering the possible values of the multiple parameters in supersymmetric theories) using random choices for mass matrix entries, allowing for CKM-like mixing of the quarks in the Dirac mass sector of the leptons.

The results were rather astonishing. As Fig. 1, taken from that work shows, Cabibbo mixing is favored between the first two "generations" and even larger mixing is highly probable between the second two, depending upon how extreme the third generation differs in mass. (Note that we did not have the temerity to consider an extreme as radical as actually occurs in the quark sector.) The mixing between the first and third "generations" is smaller, but non-vanishing. At the time, the only potential evidence for neutrino mixing was the intermediate result of the Davis experiment,[10] which was considered highly suspect, although it was later confirmed quite precisely.[11] We used our result to support mounting of neutrino oscillation experiments, saying that the mixing might well be large, although we could not predict the scale of the oscillation length.

3. Weyl Spinors

Since active neutrinos have only two basic states (as opposed to the four of a Dirac bispinor), they can be efficiently described in terms of Weyl spinors. We present the Lagrangian, equations of motion, and solutions for massive Weyl spinors, then show the relation to Majorana and Dirac constructs.

3.1. *Lagrangian density for massive Weyl spinors*

Let the Grassman-valued field variable, ϕ, represent a left-chiral $(\frac{1}{2}, 0)$ irrep of the Lorentz Group. Then

$$\mathcal{L}_L = \frac{1}{2}\phi^\dagger \sigma^\mu \overleftrightarrow{\partial}_\mu \phi + \frac{1}{2} im \left(\phi^T \sigma^2 \phi + \phi^\dagger \sigma^2 \phi^*\right) \tag{1}$$

where $\overleftrightarrow{\partial} \equiv \overrightarrow{\partial} - \overleftarrow{\partial}$ and $\sigma^\mu = (1, \sigma^i)$ with σ^i the Pauli matrices. Under a Lorentz transformation with parameters, ω_μ,

$$\phi \to e^{-\frac{i}{2}(\sigma^\mu \omega_\mu)}\phi. \tag{2}$$

For a right-chiral $(0, \frac{1}{2})$ irrep of the Lorentz Group, we need only make the substitution:

$$\mathcal{L}_R : \sigma^\mu \to \bar{\sigma}^\mu = (1, -\sigma^i) \tag{3}$$

in Eq. 1 to acquire the relevant Lagrangian.

Because mass terms must couple left-chiral $(\frac{1}{2}, 0)$ and right-chiral $(0, \frac{1}{2})$ irreps of the Lorentz Group, it is apparent that

$$\chi = \sigma^2 \phi^* \tag{4}$$

must be in a right-chiral $(0, \frac{1}{2})$ irrep, as can be seen by applying the boost in Eq. 2 to ϕ irrep and then applying the commutation rules for the Pauli matrices to the construction in Eq. 4. This will be relevant shortly.

3.2. *Equations of motion and form of solutions*

Writing ϕ out explicitly as a two component column spinor,

$$\phi = \begin{pmatrix} \phi_1 \\ \phi_2 \end{pmatrix} \tag{5}$$

the equations of motion for the components become

$$\begin{aligned} \partial_t \phi_1 - \partial_z \phi_1 - (\partial_x - \imath \partial_y)\phi_2 &= -m\phi_2^* \\ \partial_t \phi_2 + \partial_z \phi_2 - (\partial_x + \imath \partial_y)\phi_1 &= +m\phi_1^*. \end{aligned} \tag{6}$$

Defining $\theta = Et - \vec{p} \cdot \vec{x}$ and $p_\pm = p_x \pm \imath p_y$, we find the complex conjugate pair of solutions, ϕ_- and $\phi_+ = \phi_-^*$ to have the form

$$\phi_- = \begin{pmatrix} Fe^{-\imath\theta} \\ -\frac{p_+}{E-p_z} Fe^{-\imath\theta} - \imath \frac{m}{E-p_z} F^* e^{+\imath\theta} \end{pmatrix} \tag{7}$$

where F is a Grassman-valued constant.

3.3. *Majorana and Dirac bispinors*

A Majorana bispinor is simply a redundant representation of the Weyl spinor above. In the Wigner-Weyl representation for the bispinor, we make use of the transformation in Eq. 4, to construct

$$\Psi_M = \begin{pmatrix} \phi \\ e^{\imath\eta}\sigma^2\phi^* \end{pmatrix} \tag{8}$$

where the phase η can be chosen as 0, $\pm\pi/2$ or π for later convenience. The field Ψ_M has a Lagrangian that can be put into Dirac form with mass m.

To construct a Dirac bispinor, two independent $(\frac{1}{2}, 0)$ irreps must be invoked, which we labe suggestively as a, or active neutrino in the SM and s, for sterile neutrino in the SM. (Except for the $U(1)$ factor, these terms apply to the left- and right-chiral parts of the charged fermions as well.) Thus,

$$\Psi_D = \begin{pmatrix} a \\ -\sigma^2 a^* \end{pmatrix} + \imath \begin{pmatrix} s \\ -\sigma^2 s^* \end{pmatrix} = \Psi_a + \imath\Psi_s \tag{9}$$

where the phase choices have been made so that if Ψ_a and Ψ_s have the same mass value m, then (as can be seen from Eq. 1) $\imath\Psi_s$ has mass value $-m$ and a 45° rotation in the basis space will explicitly display m as a Dirac mass. (See Eq. 10 below.)

The rest states of two such independent spinors (see Eq. 7), with independent Grassman constants F and G, can be combined (with $F = -G$ and H the sum) to produce the familiar form of a spin-up Dirac particle in the Pauli-Dirac representation, *viz.*

$$\frac{1}{\sqrt{2}}\begin{pmatrix} 1 & 0 & 1 & 0 \\ 0 & 1 & 0 & 1 \\ -1 & 0 & 1 & 0 \\ 0 & -1 & 0 & 1 \end{pmatrix}\begin{pmatrix} He^{-\imath mt} \\ 0 \\ He^{-\imath mt} \\ 0 \end{pmatrix} = \begin{pmatrix} Ae^{-\imath mt} \\ 0 \\ 0 \\ 0 \end{pmatrix} \tag{10}$$

where again, F, G, H and A are all Grassman-valued constants.

4. Beyond the Simple See-Saw: Reduced Rank

The see-saw mechanism, as originally invoked, assumed, for simplicity, that the Majorana mass matrix of the right-chiral fields, represented as left-chiral but sterile neutrinos, is proportional to the unit matrix. Since then, many different "textures" for mass matrices of the fundamental fermions have been conjectured. We[12] (and others[13]) have studied the possibility that the rank of this so-called right-handed mass matrix is less than three.

With quark-lepton symmetry, the mass matrix structure consists of four three-by-three blocks: The $(1,1)$ block describes the Majorana masses of the active neutrinos and must vanish in the absence of a triplet Higgs field. The $(1,2)$ and $(2,1)$ blocks describe Dirac mass terms (m) that couple the active and sterile Weyl spinor neutrino fields. If we take them to be diagonal for the moment, this defines the flavor of each sterile neutrino field as a partners of a particular active neutrino. Finally, the $(2,2)$ block describes the Majorana masses of the sterile neutrinos. In this basis, we can initially, for a rank one sterile mass matrix, set all of the entries to zero except for the $(3,3)$ element of the$(2,2)$ block, which we label M.

These alignments are unrealistic, of course, so we carry out two sets of rotations: One corresponds to moving the vector $(0,0,M)$ in the sterile "flavor" space away from the "3" axis with the standard angles, θ (from the 3-axis) and ϕ (in the 1-2 plane). In addition, we allow for CKM-like mixing in the Dirac mass matrix sector. Absent CP-violation, this accounts for all of the possible mixings among the six fields and their corresponding particle states.

We found in this system, that there is a wide range of parameters over which the mass eigenstates form into a very light, mostly active neutrino as in the conventional see-saw, a very heavy (under the assumption that $M >> m$) mostly sterile neutrino, again conventionally, and two pairs of "pseudo-Dirac" neutrinos, the the sense of Wolfenstein.[14] The resulting mixing amplitudes among the active neutrinos are very large to maximal.

4.1. *Analytic analysis of two flavor case*

In order to understand this better, we solve analytically the simpler two flavor case. The eigen equation for this system is, after rotation from exact alignment (and ignoring the analog of CKM mixing),

$$\mu_i \Phi_i = \begin{pmatrix} 0 & 0 & m_1 & 0 \\ 0 & 0 & 0 & m_3 \\ m_1 & 0 & Ms^2 & Mcs \\ 0 & m_3 & Mcs & Mc^2 \end{pmatrix} \begin{pmatrix} \alpha_i \\ \beta_i \\ \gamma_i \\ \delta_i \end{pmatrix} \tag{11}$$

where the Φ_i are four component column vectors with entries as indicated on the RHS of the equation, and $s = sin\theta$ and $c = cos\theta$ where θ is the misalignment angle between active and sterile flavor spaces.

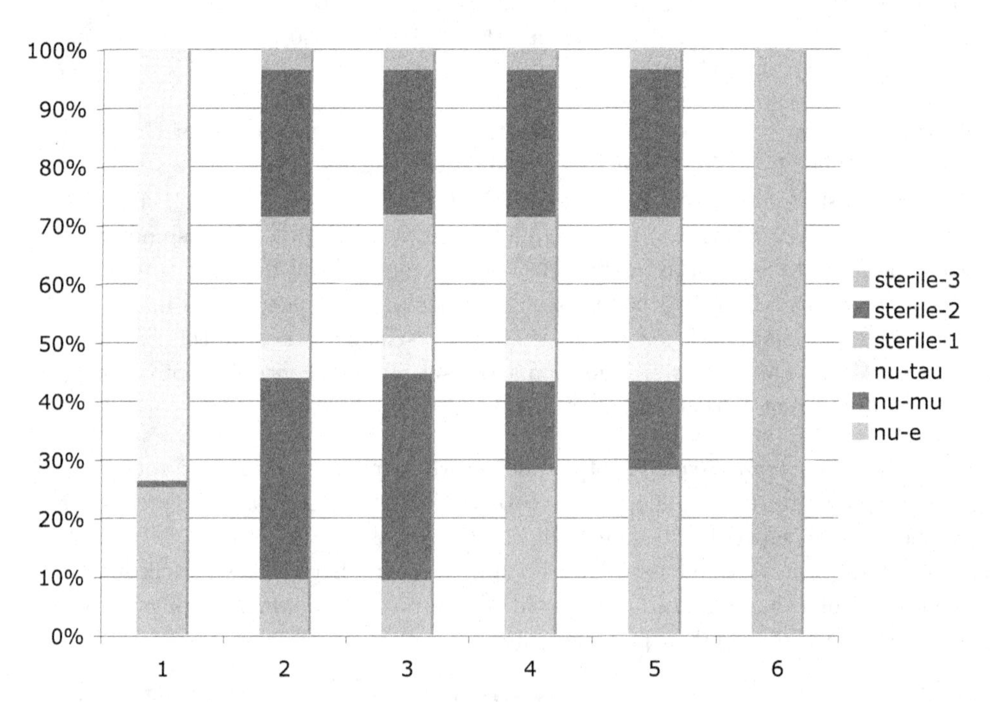

Fig. 2. Three flavor example with parameters chosen to demonstrate large flavor mixing and two pseudo-Dirac pairs ([2,3] and [4,5]). The bands indicate the relative amplitudes of each active and sterile flavor in each eigenmass state enumerated along the baseline.

McKellar showed that the eigenvalues of this system are

$$
\begin{aligned}
\mu_1 &= +m_0 - \frac{a^2}{M} - \frac{a^2}{m_0 M^2}(m_0^2 - \frac{a^2}{2} - b^2) \\
\mu_2 &= -m_0 - \frac{a^2}{M} + \frac{a^2}{m_0 M^2}(m_0^2 - \frac{a^2}{2} - b^2) \\
\mu_3 &= -\frac{b^2}{M} + \mathcal{O}(M^{-3}) \\
\mu_4 &= M + \frac{2a^2 + b^2}{M} + \mathcal{O}(M^{-3})
\end{aligned}
\tag{12}
$$

where

$$
\begin{aligned}
m_0^2 &= m_1^2 \cos^2\theta + m_3^2 \sin^2\theta \\
a &= \frac{(m_1^2 - m_3^2)\cos\theta \sin\theta}{m_0\sqrt{2}} \\
b &= \frac{m_1 m_3}{m_0}.
\end{aligned}
\tag{13}
$$

Note the low and high mass see-saw pair, μ_3 and μ_4, and the psuedo-Dirac pair, μ_1 and μ_2, which would form a single Dirac neutrino to this order if it happened that $a = 0$.

The eigenvector components for these solutions satisfy

$$
\frac{\alpha_1}{\beta_1} = \frac{\alpha_2}{\beta_2} = \frac{\beta_1}{\alpha_1} = \frac{m_1}{m_3}\cot\theta \tag{14}
$$

and

$$
\frac{\gamma_i}{\alpha_i} = \frac{\mu_i}{m_1} \quad ; \quad \frac{\delta_i}{\beta_i} = \frac{\mu_i}{m_3}. \tag{15}
$$

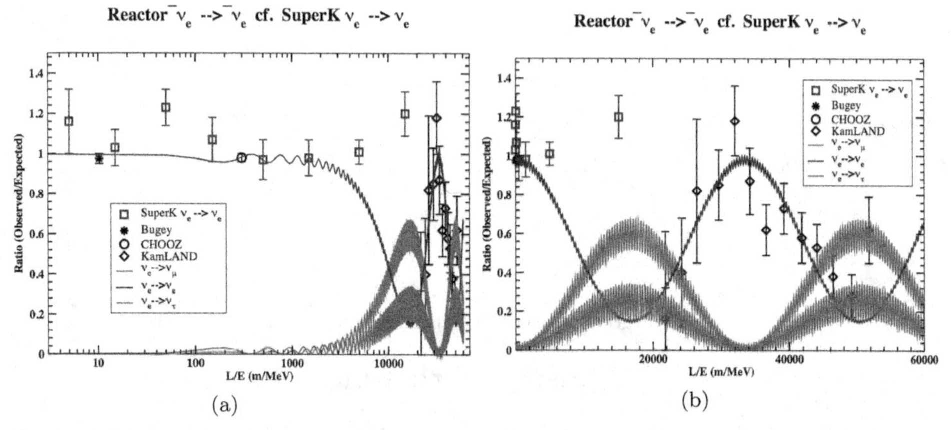

Fig. 3. Electron antineutrino disappearance vs. ratio of distance from source divided by neutrino energy compared with experimental data. (a) Logarithmic plot. (b) Linear plot.

Hence for relatively large M and small θ, eigenstates 3 and 4 are almost purely active and sterile respectively, while $\mu_1 \sim \mu_2 \sim m_1$ and the states with these two eigenvalues will have large components of both active flavor states when

$$\frac{m_3}{m_1} \sim \cot\theta$$

that is, the terms in Eq. 14 are of order one.

4.2. *Results for three flavor case*

An analytic demonstration is not so easy to provide in the three flavor case, but a similar result, with two pseudo-Dirac pairs and large flavor mixing, is shown in Fig. 2 as an example case.

In Fig. 3(a), we show the result of a specific choice of parameters for electron antineutrino disappearance, which is consistent with the results of atmospheric and reactor experiments. Within uncertainties, the last two high points of the atmospheric

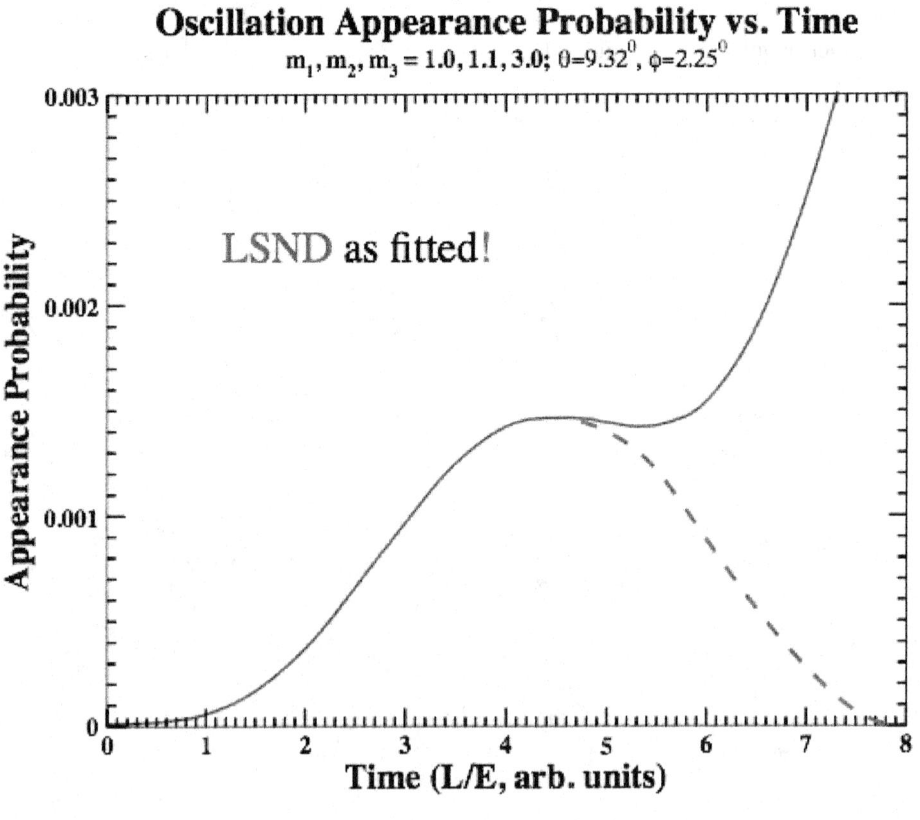

Fig. 4. Comparison of electron neutrino appearance vs. ratio of distance from source over neutrino energy as predicted by our lower rank, six-channel mixing model and the functional form (dashed line) assumed in the fit made by the experimental group.

(SuperKamiokande[15]) experiment are consistent in our parametrization with feed-through into electron neutrinos from the disappearance of muon neutrinos, also seen in that experiment. Fig. 3(b) focuses in detail on the region studied by the KamLAND experiment.[16]

In Fig. 4, we show the shape of the electron neutrino appearance function appropriate to the LSND experiment[17] and contrast this with the shape of the fitting function actually used (with the dashed extension corresponding to the simple sinusoidal function of two-channel mixing). This demonstrates that incorrect parameters may be extracted from experiments by not fitting directly to the full panoply of possibilities allowed by the three known flavors of neutrinos.

Finally, in Fig. 5, we show an example of how finite resolution, particularly in the neutrino energy, affects the oscillation pattern that is observed. The rapid oscillation between the initial muon neutrino and strongly mixed tau neutrino is smeared into

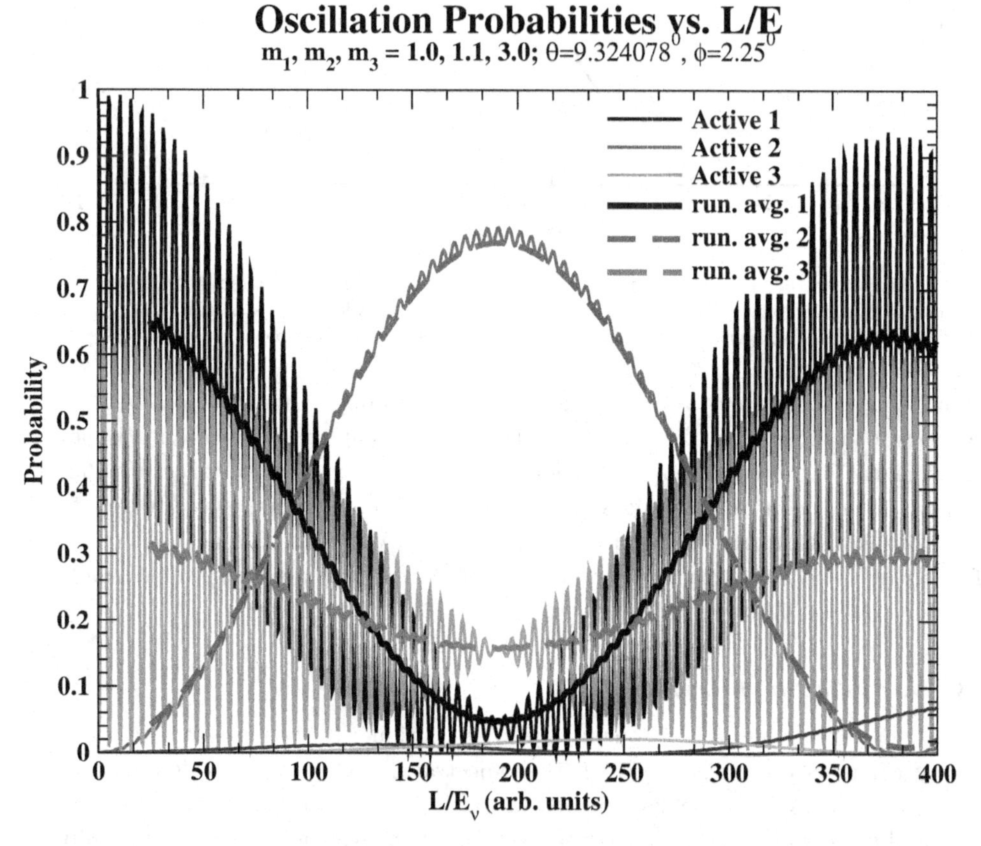

Fig. 5. Effect of finite resolution on disappearance and appearance as a function of time in the neutrino rest frame (or equivalently, ratio of distance from source to neutrino energy). The average of the disappearance of the initial neutrino flavor has approximately the expected shape for a much longer wavelength mixing than is seen at high resolution. See text for more discussion.

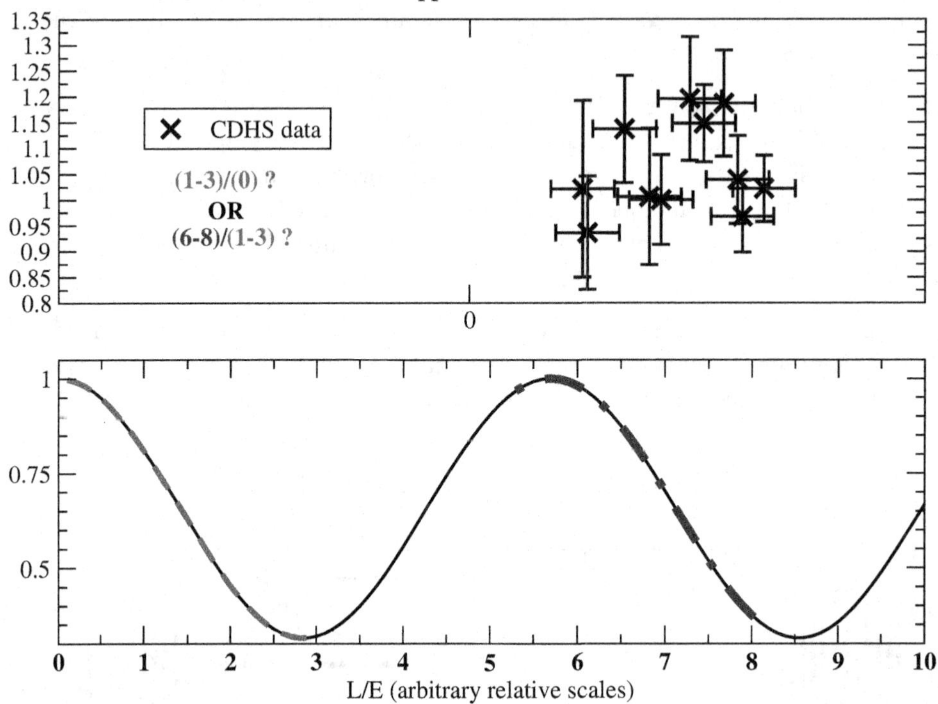

Fig. 6. Illustration of how the distance-corrected ratio of events in a far detector to those in a near detector can exceed unity if a very short wavelength oscillation affects he rate in the near detector and comparison with CDHS results.[18] See text for more discussion.

a much longer wavelength average muon neutrino disappearance. Without explicit detection of tau neutrinos, it is difficult to discern that their appearance is not following the expected, simple sinusoidal two channel appearance structure, but rather is almost constant, falling slightly while electron neutrinos actually appear in a growing fashion. Fig. 4 is a magnification of a very small region near the origin of this plot , and it is with the parameters noted here that the exceptional points referred to in Fig. 3 are explained.

4.3. *One detector or two?*

A final note in passing: A vigorous debate that continues in the experimental community concerns the question of whether spectral distortion or two detectors at different distances from the same neutrino source affords the better means to observe and study neutrino oscillation phenomena. The latter is, of course, generally more expensive, so one might be inclined to think it is also more valuable. A curious result from the CDHS experiment, however, demonstrates that one must first be certain that the near detector is so close that no oscillations at all have taken place by the time the beam arrives at that detector. In the CDHS results reported,[18] the flux in the farther detector is generally larger than in the near detector. Since this violates unitarity, it allows for a very stringent limit on neutrino disappearance. However, as we illustrate in Fig. 6, the excess can also be due to the near detector reacting to the first wave of oscillation disappearance, with the far detector appearing to have a larger (L^2 corrected) flux by being at a slightly different phase in the oscillation wave.

5. Conclusion

We draw several conclusions from the above, not all of them warranted.

- There may well be 5 independent neutrino mass differences that must be fit to neutrino oscillation experimental data.
- Analyses of oscillation data in terms of 2×2 mixing can miss significant physics and even lead to extraction of invalid parameter values.
- A global, multichannel analysis is essential before firm conclusions can be reliably drawn regarding neutrino masses and mixing parameters.
- With the ∼eV mass scales we have examined, neutrinos can contribute significantly to the Dark Matter in the Universe.

This paper includes work done with Jerry Stephenson and Bruce McKellar over many years. It has been our great pleasure to work with Bruce and we hope that, as soon as he retires, he will have a lot more time to work with us!

Acknowledgments

This work was carried out under the auspices of the National Nuclear Security Administration of the U.S. Department of Energy at Los Alamos National Laboratory under Contract No. DE-AC52-06NA25396.

References

1. Howard Georgi and S. L. Glashow, *Phys. Rev. Lett.* **32**, 438 (1974).
2. J. -E. Augustin *et al.*, *Phys. Rev. Lett.* **33**, 1406 (1974); J. J. Aubert *et al.*, *Phys. Rev. Lett.* **33**, 1404 (1974).
3. M. L. Perl *et al.*, *Phys. Rev. Lett.* **35**, 1489 (1975).
4. S. W. Herb *et al.*, *Phys. Rev. Lett.* **39**, 252 (1977) .
5. F. Abe *et al.* CDF Collaboration, *Phys. Rev. Lett.* **74**, 2626 (1995); S. Abachi *et al.* D0 Collaboration, *Phys. Rev. Lett.* **74**, 2632 (1995).
6. M. Gell-Mann, P. Ramond and R. Slansky, in *Supergravity*, Proceedings of the Workshop, Stony Brook, New York, 1979, ed. by P. van Nieuwenhuizen and D. Freedman (North-Holland, Amsterdam, 1979), p. 315; T. Yanagida, in *Proceedings of the Workshop on the Unified Theories and Baryon Number in the Universe*, Tsukuba, Japan, 1979, edited by O. Sawada and A. Sugamoto (KEK Report No. 79-18, Tsukuba, 1979), p.95; R. N. Mohapatra and G. Senjanovic, *Phys. Rev. Lett.* **44**, 912 (1980); S. L. Glashow, in *Quarks and Leptons*, Cargese (July 9-29, 1979), eds. M. Levy *et al.* (Plenum, New York, 1980), p. 707.

7. N. Cabibbo, *Phys. Rev. Lett.* **10**, 531 (1963); M. Kobayashi and T. Maskawa, *Prog. Theor. Phys.* **49**, 652 (1973).
8. B. Pontecorvo, *Zh. Eksp. Teor. Fiz.* **33**, 549(1957); **34**, 247 (1958).
9. T. Goldman and G. J. Stephenson, Jr., *Phys. Rev.* **D24**, 236 (1981).
10. R. Davis, Jr., D. S. Harmer, and K. C. Hoffman, *Phys. Rev. Lett.* **20**, 1205 (1968).
11. Q. R. Ahmad *et al.*, *Phys. Rev. Lett.* **89**, 011301(2002); **87**, 071301 (2001); S. Fukuda *et al. Phys. Lett. B* **539**, 179 (2002).
12. G. J. Stephenson, Jr., T. Goldman, B. H. J. McKellar and M. Garbutt, *Int. J. Mod. Phys.* **A20**, 6373 (2005); [hep-ph/0404015].
13. K. S. Babu, B. Dutta and R. N. Mohapatra, *Phys. Rev. D* **67**, 076006 (2003); hep-ph/0211068.
14. L. Wolfenstein, *Phys. Lett.* **B107**, 77 (1981); *Nucl. Phys.* **B186**, 147 (1981).
15. Y. Ashie *et al.* Super-Kamiokande Collaboration, *Phys. Rev. D* **71**, 112005 (2005); Y. Fukuda *et al.*, *Phys. Rev. Lett.* **81**, 1562 (1998).
16. T. Araki *et al.*, *Phys. Rev. Lett.* **94**, 081801 (2005).
17. C. Athanassopoulos *et al.*, *Phys. Rev. Lett.* **77**, 3082 (1996); A. Aguilar *et al.*, *Phys. Rev. D***64**, 112007 (2001).
18. F. Dydak *et al.*, *Phys. Lett.* **134B**, 281 (1984).

PARTICLE PHYSICS IN CONDENSED MATTER

CHRIS HAMER*
School of Physics, University of New South Wales
Sydney NSW 2052, Australia
**E-mail: C.Hamer@unsw.edu.au*

We highlight the surprising similarities between particle physics and condensed matter physics which have been emerging from recent advances in experimantal and theoretical techniques.

Keywords: Particle physics; condensed matter; pseudoparticles; confinement.

1. Introduction

Girish Joshi has spent his research career in theoretical particle physics, while Bruce McKellar moved from nuclear into particle physics. My research career, on the other hand, started in particle physics and moved sideways into the domain of theoretical condensed matter physics. I thought it might be interesting on this auspicious occasion to trace out some of the surprising similarities between many aspects of particle and condensed matter physics.
I will start with a brief review of quantum lattice models in condensed matter physics, the concept of quasiparticles, and the phenomena of quantum phase transitions. I will discuss the energy spectra and dispersion relations of the quasiparticles. The analogue of "deep inelastic scattering" will be discussed, and also the occurrence of bound states and "confinement" in condensed matter physics. Finally. some conclusions will be drawn.

2. Quantum Lattice Models

Our simplified theoretical models of crystalline solids generally begin with a lattice (Figure 1), whose sites i represent the atoms (or possibly molecules) of the crystal. Some of the standard lattice models are:
(i) The XXZ Heisenberg model:

$$H = J \sum_{<ij>} [S_i^z S_j^z + \Delta(S_i^x S_j^x + S_i^y s_j^y)] \tag{1}$$

where the $\mathbf{S}_i$ represent atomic spins, and the $< ij >$ denote nearest-neighbour pairs of sites, between which we postulate a spin-spin interaction. This is a very simple

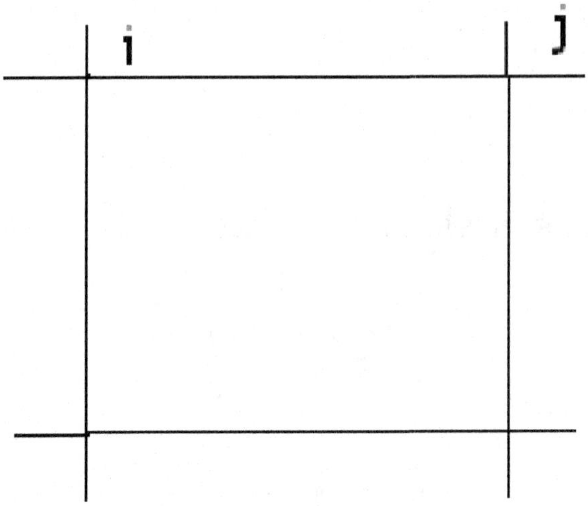

Fig. 1. Lattice representation of a two-dimensional crystal.

model of a ferromagnetic or antiferromagnetic material (depending on the sign of the coupling J). It is called a 'quantum' lattice model because it contains operators (the S_i^α) which do not all commute.
(ii) The 'transverse' Ising model:

$$H = \sum_I (1 - \sigma_i^z) - \lambda \sum_{<ij>} \sigma_i^x \sigma_j^x \tag{2}$$

where the σ_I^α are Pauli matrices acting on S=1/2 spin variables at each site. This has various physical applications, and is perhaps the simplest paradigm of all quantum lattice models.
(iii) The Hubbard model:

$$H = -t \sum_{<ij>} (c_{I\sigma}^\dagger c_{j\sigma} + U \sum_i n_{i\uparrow} n_{i\downarrow} \tag{3}$$

which is the prototype model for conductors or superconductors, where the t term is a fermion "hopping" term, and the U term is an on-site approximation to the Coulomb repulsion between fermions.
The partition function

$$Z = \sum_n e^{-\beta E_n} \tag{4}$$

is dominated at low temperatures by the ground state and the low-energy excitations. For instance, for the transverse Ising model in one space dimension at $\lambda \ll 1$, the ground state is an ordered ferromagnetic state

$$|0> = |\uparrow\uparrow \cdots \uparrow> \tag{5}$$

with spins up along the z axis at all sites i. The lowest excitation is a single flipped spin

$$|1> = |\uparrow\downarrow\uparrow \cdots \uparrow> \tag{6}$$

which may travel through the lattice: a 'magnon' excitation. The excitation may carry energy and momentum just like a particle in free space, and is called a *'quasi-particle'*.
For $\lambda \gg 1$, however, there are two degenerate ground states. When we quantize spins along the x axis, they are

$$|0> = |\uparrow\uparrow \cdots \uparrow> \tag{7}$$

and

$$|0>' = |\downarrow\downarrow \cdots \downarrow> . \tag{8}$$

The lowest excited state is

$$|0> = |\uparrow\uparrow\downarrow\downarrow \cdots \downarrow>, \tag{9}$$

a 'domain wall' between spin up and spin down domains, called a 'spinon' excitation. Many other quasiparticle or pseudoparticle excitations may occur in condensed matter physics, including

$$\begin{array}{cc} electrons/holes & visons \\ excitons & rotons \\ polarons & anyons \\ \cdots & \end{array} \tag{10}$$

- a whole zoo of excitations, much as in particle physics! The energy scales of these excitations are typically eV or meV (millielectron volts), rather than MeV.

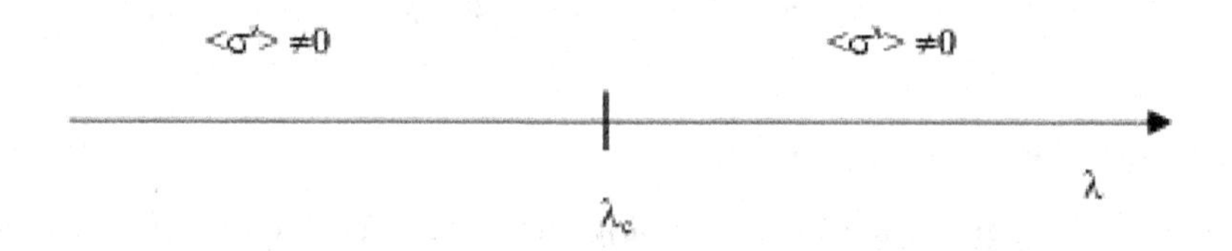

Fig. 2. Phase diagram for the transverse Ising model.

3. Quantum Phase Transitions

'Quantum' phase transitions are transitions in a lattice model driven by changes in the coupling constants, rather than by temperature, say. For instance, in the transverse Ising model at fixed temperature $T = 0$, there is a transition between the ordered states mentioned above, driven by the coupling λ (Figure 2). The order parameter in the small λ phase is $< \sigma^z >_0$, and in the large λ phase $< \sigma^x >_0$,

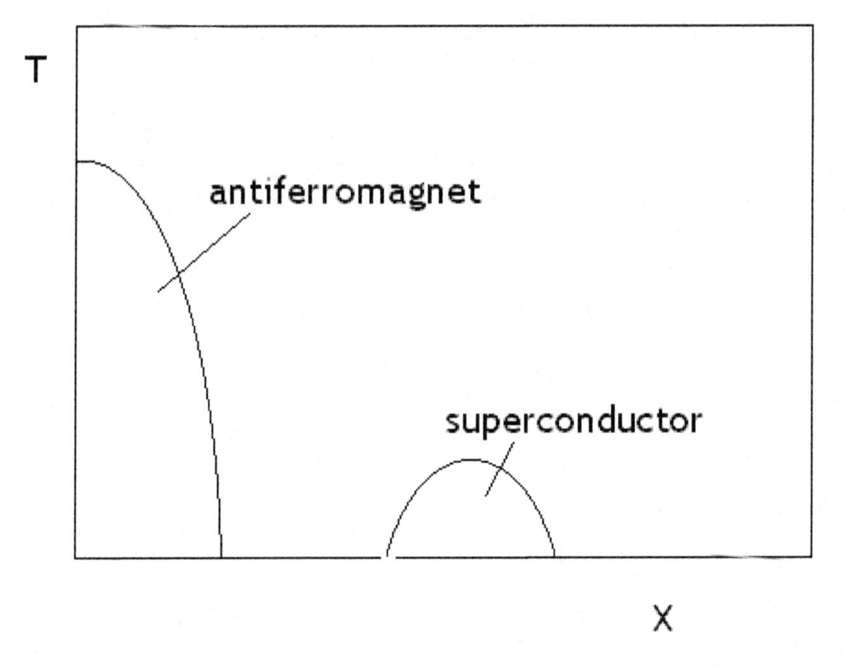

Fig. 3. Schematic phase diagram for a high-temperature superconductor.

and there is a phase transition between them at some critical λ_c. For the quantum model in d dimensions, the transition turns out to be in the same universality class as the classical Ising model in (d+1) space dimensions.
For superconductors, the superconducting state may also be associated with a quantum phase transition at $T = 0$, for example the high-temperature superconductors (Figure 3), where the superconducting state is induced by changing the doping fraction, x, or heavy-fermion materials where it is induced by changing the pressure.

4. Quasiparticle Energy Spectra

In relativistic particle physics, the energy of a free particle is given by the familiar form

$$E = \sqrt{p^2c^2 + m^2c^4}. \tag{11}$$

In condensed matter, the dispersion relation may be much more interesting! Figure 4 shows the dispersion relation for a single magnon in the Heisenberg model on a square lattice.[1] As in particle physics, *Goldstone bosons* can appear due to spontaneous symmetry breaking. They may have linear dispersion relations (e.g. antiferromagnets), as illustrated in Figure 4 near $\mathbf{k} = (0,0)$, or quadratic dispersion relations (e.g. ferromagnets).

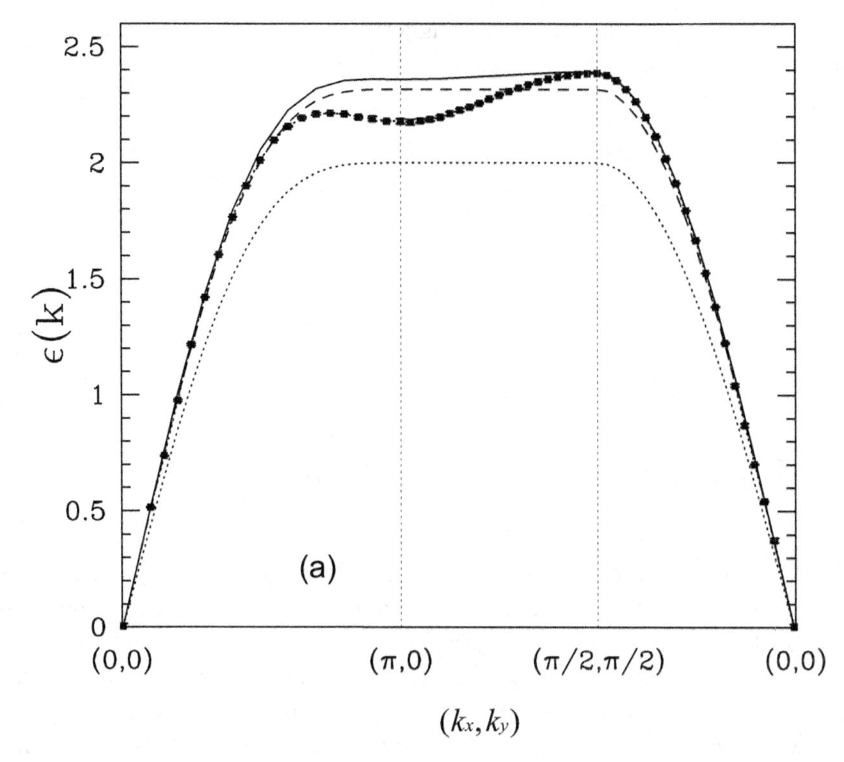

Fig. 4. The one-magnon excitation spectrum $\epsilon(\mathbf{k})$ along high-symmetry cuts in the Brillouin zone for the Heisenberg antiferromagnet on the square lattice.[1] Also shown are the results of first-order (dotted line), second-order (dashed line), and third-order (solid line) spin-wave theory.

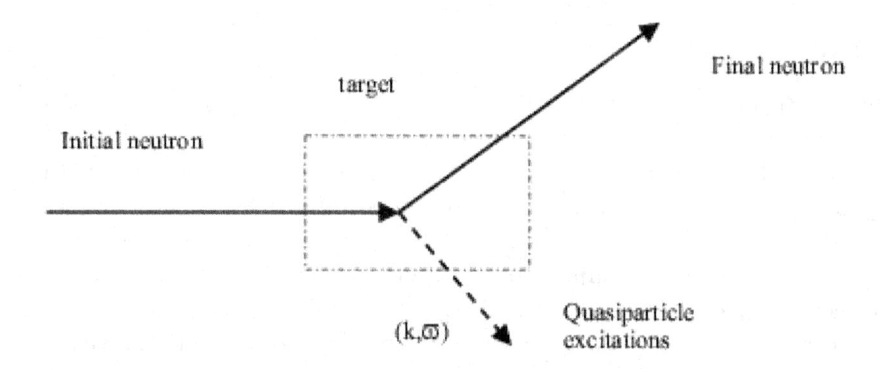

Fig. 5. A neutron scattering from a target material, inducing quasiparticle excitations.

5. "Deep Inelastic" Scattering

Scattering experiments may be used, just as in particle physics, to study the structure and dynamics of a target material in condensed matter. Figure 5 shows how a neutron scattering experiment can be used to create quasiparticle excitations in a material. The neutrons couple to *spin* via magnetic interactions, so the cross-section is proportional to the spin "structure factor"

$$\frac{d\sigma}{d\Omega} \propto S(\mathbf{k},\omega) \tag{12}$$

where

$$S^{\alpha\beta}(\mathbf{k},\omega) = \frac{1}{2\pi N}\sum_{i,j}\int_{-\infty}^{\infty} e^{i(\omega t+\mathbf{k}.(\mathbf{r_i}-\mathbf{r_j}))} < S^{\alpha}_{j}(t)S^{\beta}_{i}(0) >_0 . \tag{13}$$

Insert a complete set of states Λ, then

$$S^{\alpha\beta}(\mathbf{k},\omega) = \sum_{\Lambda}\delta(\omega + E_0 - E_\Lambda(\mathbf{k}))S^{\alpha\beta}_{\Lambda}(\mathbf{k}) \tag{14}$$

where $S^{\alpha\beta}_{\Lambda}(\mathbf{k})$ is the "spectral weight" of state Λ. Neutron scattering experiments can measure these quantities directly, see for example Xu *et al.*,[2] where the sinusoidal trace of the single magnon excitation is clearly seen, or Tennant *et al.*,[3] a later experiment in which the contribution of two-magnon states can also be seen. Note that *bound states* may occur in these systems, just as in particle physics. Figure 6 shows a theoretical calculation of the 2-particle spectrum for the alternating Heisenberg chain.[4] Near $k = \pi/2$, clear evidence is seen of singlet bound states S_1 and S_2 and triplet bound states T_1 and T_2 below the continuum, as well as quintuplet antibound states Q_1 and Q_2 above the continuum.

6. Confinement

Quasiparticle "confinement" may occur in condensed matter physics, just as in particle physics. One case discussed by Affleck *et al.*[5] is that of the alternating spin-1/2 Heisenberg chain (Figure 7a)), a Heisenberg antiferromagnet with alternating couplings J and δJ. At $\delta = 0$, the ground state consists of spin $S = 0$ "dimers", with $\mathbf{S_1}\cdot\mathbf{S_2} = -3/4$ on each pair of sites with coupling J (Fig. 7b)). A 'spinon' excitation (unpaired spin) can occur in conjunction with an antispinon, as illustrated in Fig. 7c), but the energy of this configuration is proportional to the distance between them L, because the dimers in between are on the 'wrong' (weakly coupled) pairs of sites: *Linear confinement.* The limit $\delta = 1$ corresponds to the uniform Heisenberg chain, where the spinons become deconfined.

One of the hot topics in theoretical condensed matter at the moment is the search for a 'deconfined quantum critical point' where fractional excitations such as spinons may become deconfined in two-dimensional models.[6] It is predicted that these will belong to new universality classes, beyond the traditional Landau-Ginzburg-Wilson paradigm.

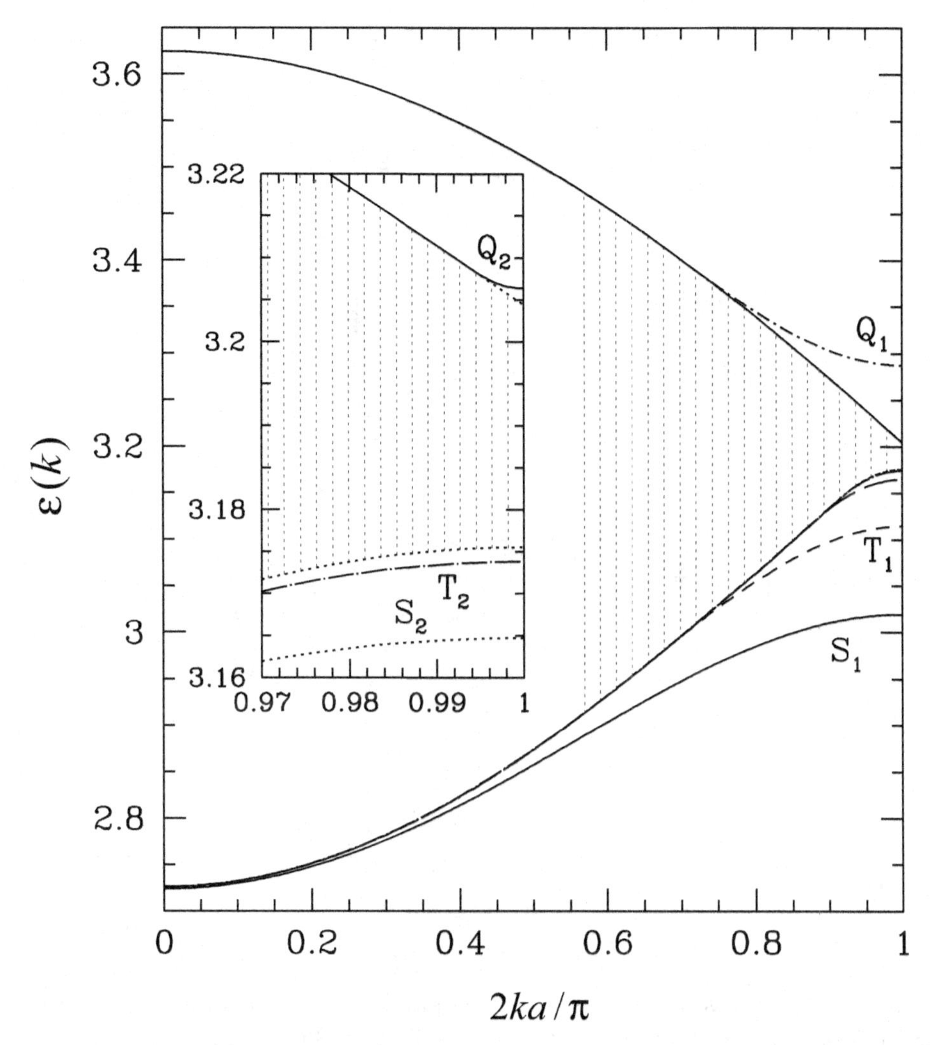

Fig. 6. The two-particle excitation spectrum of the $J_1 - J_2 - \delta$ chain with $\alpha = 0$ and $\delta = 0.6$. Note the bound and antibound states lying outside the two-particle continuum (shaded).[4]

7. Conclusions

As we have sketched in these discussions, the occurrence of (quasi)particles is universal in condensed matter systems, as well as in particle physics. They are intrinsically connected with the structure and dynamics of the system. They can display a wide variety of dispersion relations. They can form bound states, and undergo scattering processes just like relativistic particles.

We have discussed analogues of deep inelastic scattering and confinement phenomena in condensed matter physics. One of the current issues in the field is the search for deconfined fractional excitations and non-Landau-Ginzburg-Wilson phase transitions in two space dimensions at 'deconfined quantum critical points'.

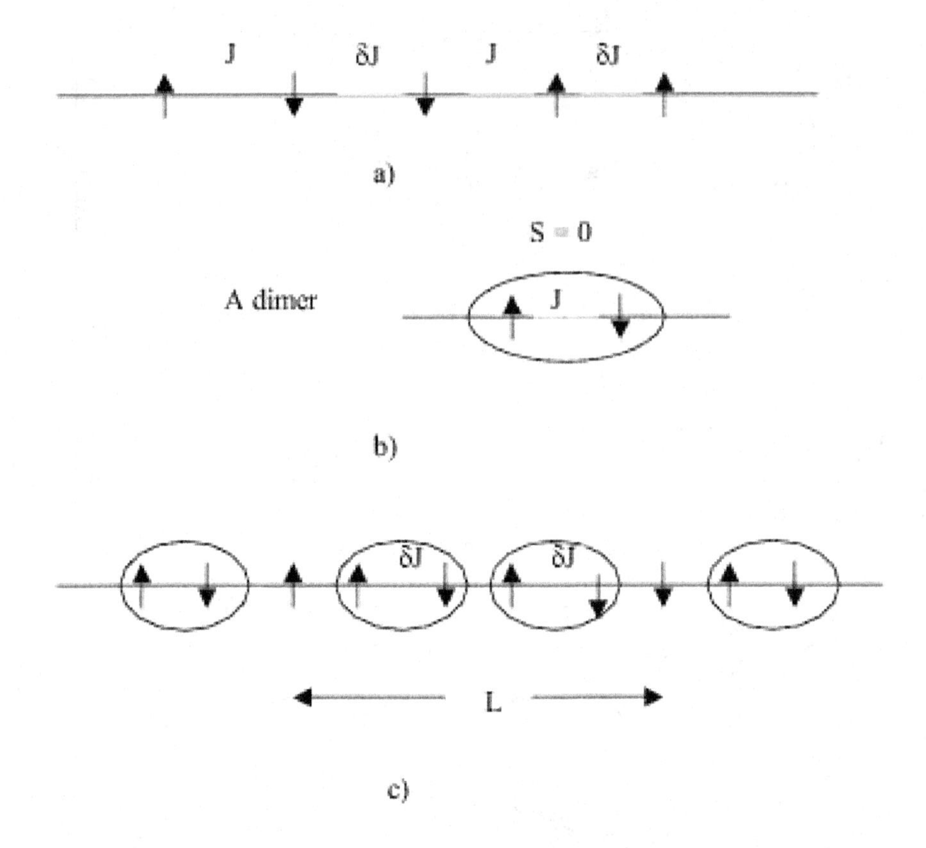

Fig. 7. a) The alternating Heisenberg chain; b) a spin-0 dimerized pair of sites; c) a spinon-antispinon excitation.

There are no analogues in condensed matter physics, as yet, of the giant particle accelerators and detector systems in particle physics. But there have been some recent developments pointing in this direction. These include experiments on single-electron transport in nanowires, and work on injector systems for spin excitations in "spintronics". There is likely to be much more work on spintronics in the future aimed at generating, storing and controlling spin excitations, with the goal of developing new computing systems.

References

1. W. Zheng, J. Oitmaa and C.J. Hamer, Phys. Rev. **B71**, 184440 (2005).
2. G. Xu, C. Broholm, D.H. Reich and M.A. Adams, Phys. Rev. Lett. **84**, 4465 (2000).
3. D.A. Tennant, C. Broholm, D.H. Reich, S.E. Nagler, G.E. Granroth, T. Barnes, K. Damle, G. Xu, Y. Chen and B.C. Sales, Phys. Rev. **B67**, 054414 (2003).
4. J. Oitmaa, C.J. Hamer and W. Zheng, *Series Expansion Methods for Strongly Interacting Lattice Models* (Cambridge University Press, 2006), p. 143.

5. E.S. Sorenson, I. Affleck, D. Augier and D. Poilblanc, Phys. Rev. **B58**, R14701 (1998); I. Affleck, in *Dynamical Properties of Unconventional Magnetic Systems* (NATO ASI, Geilo, Norway, 1997).
6. T. Senthil, A. Vishwanath, L. Balents, S. Sachdev and M.P.A. Fisher, *Science* **303**, 1490 (2004); Phys. Rev. **B70**, 144407 (2004).

CP VIOLATION

XIAO-GANG HE*

Department of Physics and Center for Theoretical Sciences, National Taiwan University, Taipei, Taiwan

In this contribution to the Festschrift in honor of Bruce McKellar and Girish Joshi, I discuss a model proposed recently[1] in which the CP violating phase in the CKM matrix is identical to the phase in the Higgs potential resulting from spontaneous CP violation.

Keywords: CP violation, spontaneous symmetry breaking.

1. Introduction

It is a great pleasure to give a talk at the Festschrift in honor of Professors Bruce McKellar and Girish Joshi. I have learnt a lot from them on the topic of this talk: CP violation. I discuss a model proposed recently[a] in which the CP violating phase in the CKM matrix is identical to the phase in the Higgs potential resulting from spontaneous CP violation.

2. CP Violation

Since the discovery of parity (P) violation in weak interactions[2] by T.D. Lee and C.-N. Yang in 1956, great progress has been made in many ways. Parity violation is now understood to be due to $V - A$ current interaction in weak interactions. We have a successful theory for electroweak interactions - the Standard Model (SM).[3] More understanding on discrete space-time symmetries, parity P, time reversal T and charge conjugation C, has also been gained over the last 50 years or so. In 1964, CP was found to be violated in neutral kaon decays into two and three pions.[4] Since then CP violation has been observed in the kaon decay amplitude, and in B meson decays in recent years.[5] Now we have a very successful model for CP violation, the CKM model.[6] T violation is also established in kaon decays.[5] There is no evidence of violation for the combined symmetry CPT.

It was realized in 1973 by Kobayashi and Maskawa[6] that in the minimal SM if there is a miss-match between weak and mass eigenstates of quarks in the interaction with the weak gauge boson W and Higgs boson, it is possible to have CP violation. In the

*hexg@phys.ntu.edu.tw

[a]Now in preprint form in Ref. 1.

weak interaction basis, one can write the renormalizable W and Higgs interactions with quarks as the following

$$\mathcal{L} = -\frac{g}{\sqrt{2}}\bar{U}_L\gamma^\mu D_L W^+_\mu - \frac{1}{\sqrt{2}}(\bar{U}_L\lambda^U U_R + \bar{D}_L\lambda^d D_R)(v+h) + h.c.$$

where v is the vacuum expectation value (VEV) of the Higgs doublet H in the SM.

The quark mass matrices $M^{U,D} = \lambda^{U,D}v/\sqrt{2}$ are arbitrary $N \times N$ matrices for N generations of quarks and are, in general, not diagonal in weak basis. In the mass eigenstate basis, $M^{U,D}$ become diagonalized and the W interaction with quarks will not be diagonal with

$$\mathcal{L} = -\frac{g}{\sqrt{2}}\bar{U}\gamma^\mu V_{KM} LDW^+_\mu + h.c. - (\bar{U}\hat{M}^U U + \bar{D}\hat{M}^D D)(1+\frac{h}{v}),$$
$$V_{KM} = V_L^{U\dagger}V_L^D, M^{U,D} = V_L^{U,D}\hat{M}^{U,D}V_R^{U,D},$$

where $\hat{M}^{U,D}$ are diagonal matrices. V_{KM} is also called the CKM matrix. $V_{CKM} = V_{KM}$ in general has $N(N-1)/2$ mixing angles and $(N-1)(N-2)/2$ phases. A non-zero value for the sine of the phases lead to CP violation. The minimal number of generations for CP violation is three.

With three generations of quarks, V_{CKM} can be written as

$$V_{CKM} = \begin{pmatrix} V_{ud} & V_{us} & V_{ub} \\ V_{cd} & V_{cs} & V_{cb} \\ V_{td} & V_{ts} & V_{tb} \end{pmatrix}$$
$$= \begin{pmatrix} c_{12}c_{13} & s_{12}c_{13} & s_{13}e^{-i\delta_{13}} \\ -s_{12}c_{23} - c_{12}s_{23}s_{13}e^{i\delta_{13}} & c_{12}c_{23} - s_{12}s_{23}s_{13}e^{i\delta_{13}} & s_{23}c_{13} \\ s_{12}s_{23} - c_{12}c_{23}s_{13}e^{i\delta_{13}} & -c_{12}s_{23} - s_{12}c_{23}s_{13}e^{i\delta_{13}} & c_{23}c_{13} \end{pmatrix}, \quad (1)$$

where $s_{ij} = \sin\theta_{ij}$ and $c_{ij} = \cos\theta_{ij}$, and $\gamma = \delta_{13}$. One usually uses the Wolfenstein parameterization (λ, A, ρ, η) for convenience[7] with $V_{us} = s_{12}c_{13} = \lambda$, $V_{ub} = s_{13}e^{-i\gamma} = A\lambda^3(\rho - i\eta)$, $V_{cb} = s_{23}c_{13} = A\lambda^2$.

The CKM model is very successful in describing all laboratory experimental results related to CP violation and mixing phenomena. In Fig. 1 we show the current constraints from various experiments.[8] The best fit values for the parameters are[5]

$$\lambda = 0.2272 \pm 0.0010\,, \quad A = 0.818^{+0.007}_{-0.017} \quad \rho = 0.221^{+0.064}_{-0.028}\,, \quad \eta = 0.340^{+0.017}_{-0.045}\,.$$

The CKM model, although successful, provides no explanation of where the CP violating phase comes from, which calls for more theoretical studies. There is also a hint from matter and anti-matter asymmetry in our universe that there is the need for CP violation beyond the CKM model since it gives a too small value for the observed asymmetry. The origin of CP violation is an outstanding problem of particle physics.

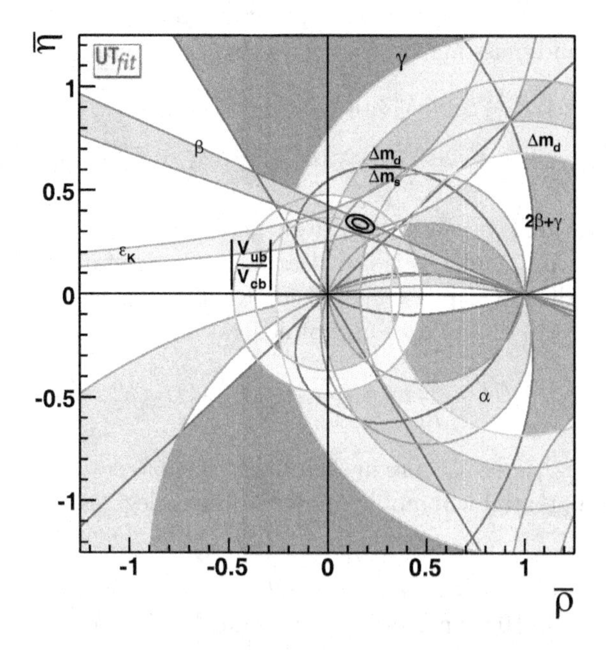

Fig. 1. Constraints on the CKM parameters.

Also in 1973, T. D. Lee[9] proposed that CP violation can come from symmetry breaking in the vacuum: spontaneous CP violation. This provides an understanding of the origin of CP violation. In the minimal SM this is, however, not possible. With more than one Higgs doublet, it can be realized. For example with two Higgs doublets, ϕ_1 and ϕ_2, the most general potential one can write down is given by[9]

$$\begin{aligned} V(\phi) = &-\lambda_1\phi_1^\dagger\phi_1 - \lambda_2\phi_2^\dagger\phi_2 - \lambda_{12}(\phi_1^\dagger\phi_2 + \phi_2^\dagger\phi_1) \\ &+A(\phi_1^\dagger\phi_1)^2 + B(\phi_2^\dagger\phi_2)^2 + C(\phi_1^\dagger\phi_1)(\phi_2^\dagger\phi_2) + \bar{C}(\phi_1^\dagger\phi_2)(\phi_2^\dagger\phi_1) \\ &+\frac{1}{2}[(\phi_1^\dagger\phi_2)(D\phi_1^\dagger\phi_2 + E\phi_1^\dagger\phi_1 + F\phi_2^\dagger\phi_2) + h.c.]. \end{aligned}$$

Writing the VEVs as $\langle\phi_1^0\rangle = \rho_1 e^{i\theta}/\sqrt{2}$, $\langle\phi_2^0\rangle = \rho_2/\sqrt{2}$, if $\theta \neq 0$ CP is spontaneously violated. One of the conditions for the Higgs potential to be minimal at the VEV is

$$\left.\frac{\partial V}{\partial\theta}\right|_{min} = (2\lambda_{12} - 4D\cos\theta - E\rho_1^2 - F\rho_2^2)\rho_1\rho_2\sin\theta = 0. \qquad (2)$$

This has a non-trivial solution for spontaneous CP violation which is given by

$$\cos\theta = \frac{1}{4D}(2\lambda_{12} - E\rho_1^2 - F\rho_2^2). \qquad (3)$$

The above two-Higgs-doublet model has complicated interactions with quarks. The Yukawa interactions and mass matrices are given by

$$\mathcal{L}_{\phi-q} = -\bar{Q}_L(\lambda_1^u\phi_1 + \lambda_2^u\phi_2)U_R - \bar{Q}_L(\lambda_1^d\tilde{\phi}_1 + \lambda_2^d\tilde{\phi}_2)D_R,$$
$$M_u = \frac{1}{\sqrt{2}}(\lambda_1^u\rho_1 e^{i\theta} + \lambda_2^u\rho_2),\ \ M_d = \frac{1}{\sqrt{2}}(\lambda_1^d\rho_1 e^{-i\theta} + \lambda_2^d\rho_2).$$

Even though $\lambda_{1,2}^{u,d}$ are real, because $M_{u,d}$ are complex one can obtain a complex V_{CKM}. However in such a model, it is not clear how the spontaneous CP violating phase θ is related to the CKM phase δ_{13}. There are also tree level flavor changing neutral currents (FCNCs) due to exchange of neutral Higgs,

$$\mathcal{L}_{neutral} = -\frac{1}{\sqrt{2}}[\bar{U}_L\hat{\lambda}_2^u U_R(h_2^0 - \frac{\rho_2}{\rho_1}h_1^0 + i\frac{\rho}{\rho_1}A) + \bar{D}_L\hat{\lambda}_2^d D_R(h_2^0 - \frac{\rho_2}{\rho_1}h_1^0 - i\frac{\rho}{\rho_1}A)],$$

$\rho = \rho_1^2 + \rho_2^2$, The hat indicates $\lambda_2^{u,d}$ are in their mass eigenbases which are in general complex. The Higgs potential will mix the physical Higgs degrees of freedom A and h_i^0.

The model can be made to be consistent with data, but CP violation is very non-CKM like in general, and Higgs masses are constrained to be large due to tree level FCNCs. To avoid FCNCs at tree level, Weinberg[10] in 1976 proposed to impose additional discrete symmetries such that only one Higgs doublet gives masses to the up and/or down quark sectors. In this case three Higgs doublets are needed to have spontaneous CP violation. In the Weinberg model the CKM matrix is real and in conflict with experimental data,[11] in particular with data for $\sin 2\beta_{eff} = 0.678\pm0.032$ from $B \to J\psi K$ decay.[5] In the Weinberg model there is no contribution from CKM sector, and the charged Higgs contribution to $\sin 2\beta_{eff}$ is less than 0.05. The Weinberg model of spontaneous CP violation is decisively ruled out by CP violation in $B \to J/\psi K$.

In the following we present a model[1] where the CP violating phase in the CKM matrix has a clear relation with the spontaneous CP violating phase in the Higgs potential by identifying these two phases to be the same up to a sign.

We start the discussion by showing that the above idea is indeed realizable. Let us consider the following Yukawa couplings with multi-Higgs doublets,

$$L_Y = \bar{Q}_L(\Gamma_{u1}\phi_1 + \Gamma_{u2}\phi_2)U_R + \bar{Q}_L\Gamma_d\tilde{\phi}_d D_R + h.c.\ , \tag{4}$$

$\tilde{\phi}_d = -i\sigma_2\phi_d^*$ and ϕ_d may be one of the $\phi_{1,2}$ or another doublet Higgs field. The Yukawa couplings $\Gamma_{u1,u2,d}$ must be real if CP is only violated spontaneously.

The Higgs doublets when expressed in terms of the component fields and their VEVs v_i are given by

$$\phi_i = e^{i\theta_i}H_i = e^{i\theta_i}\begin{pmatrix} \frac{1}{\sqrt{2}}(v_i + R_i + iA_i) \\ h_i^- \end{pmatrix}. \tag{5}$$

The quark mass terms in the Lagrangian are

$$L_m = -\bar{U}_L \left[M_{u1} e^{i\theta_1} + M_{u2} e^{i\theta_2}\right] U_R - \bar{D}_L M_d e^{-i\theta_d} D_R + h.c. , \tag{6}$$

where $M_{ui} = -\Gamma_{ui} v_i / \sqrt{2}$.

The phases θ_1 and θ_d can be absorbed by redefining the fields U_R and D_R. However, the phase difference $\delta = \theta_2 - \theta_1$ cannot be removed and it depends on the Higgs potential. A non-zero δ indicates spontaneous CP violation. Without loss of generality, we work in the basis where D_L, D_R are already in their mass eigenstates. In this basis the down quark mass matrix M_d is diagonalized, which will be indicated by $\hat{M}_d$. In general the up quark mass matrix $M_u = M_{u1} + e^{i\delta} M_{u2}$ is not diagonal. Diagonalizing M_u produces the CKM mixing matrix. One can write $\hat{M}_u = V_{CKM} M_u V_R^\dagger$. V_R is an unknown unitary matrix. A direct identification of the phase δ with the phase δ_{13} in the CKM matrix is not possible in general. There are, however, classes of mass matrices which allow such a connection. A simple example is provided by setting V_R to be the unit matrix. With this condition, $M_u = V_{CKM}^\dagger \hat{M}_u$. One then needs to show that $V_{CKM}^\dagger$ can be written as

$$V_{CKM}^\dagger = (M_{u1} + e^{i\delta} M_{u2}) \hat{M}_u^{-1}. \tag{7}$$

Expressing the CKM matrix in this form is very suggestive. If V_{CKM} can always be written as a sum of two terms with a relative phase, then the phase in the CKM matrix can be identified with the phase δ.

We now demonstrate that it is the case by using the parametrization in Eq. (1) as an example. To get as close as to the form in Eq. (7), we write the CKM matrix Eq. (1) as[5]

$$V_{CKM} = \begin{pmatrix} e^{-i\delta_{13}} & 0 & 0 \\ 0 & 1 & 0 \\ 0 & 0 & 1 \end{pmatrix} \begin{pmatrix} c_{12} c_{13} e^{i\delta_{13}} & s_{12} c_{13} e^{i\delta_{13}} & s_{13} \\ -s_{12} c_{23} - c_{12} s_{23} s_{13} e^{i\delta_{13}} & c_{12} c_{23} - s_{12} s_{23} s_{13} e^{i\delta_{13}} & s_{23} c_{13} \\ s_{12} s_{23} - c_{12} c_{23} s_{13} e^{i\delta_{13}} & -c_{12} s_{23} - s_{12} c_{23} s_{13} e^{i\delta_{13}} & c_{23} c_{13} \end{pmatrix} .$$

Absorbing the left matrix into the definition of U_L field, we have

$$M_{u1} = \begin{pmatrix} 0 & -s_{12} c_{23} & s_{12} s_{23} \\ 0 & c_{12} c_{23} & -c_{12} s_{23} \\ s_{13} & s_{23} c_{13} & c_{23} c_{13} \end{pmatrix} \hat{M}_u ,$$

$$M_{u2} = \begin{pmatrix} c_{12} c_{13} & -c_{12} s_{23} s_{13} & -c_{12} c_{23} s_{13} \\ s_{12} c_{13} & -s_{12} s_{23} s_{13} & -s_{12} c_{23} s_{13} \\ 0 & 0 & 0 \end{pmatrix} \hat{M}_u , \tag{8}$$

and $\delta = -\delta_{13}$. We therefore find that it is possible to identify the CKM phase with that resulting from spontaneous CP violation. Note that as long as the phase δ is not zero, CP violation will show up in the charged currents mediated by W exchange. The effects do not disappear even when Higgs boson masses are all set to be much higher than the W scale. Furthermore, $M_{1,2}$ are fixed in terms of the CKM matrix elements and the quark masses, as opposed to being arbitrary in general multi-Higgs models.

We comment that the solution is not unique even when V_R is set to be the unit matrix.[1] One can easily verify this by taking another parametrization for the CKM matrix, such as the original Kobayashi-Maskawa (KM) matrix.[6] More physical requirements are needed to uniquely determine the connection. The key point we would like to emphasize is that there are solutions where the phase in the CKM matrix can be identified with the phase causing spontaneous CP violation in the Higgs potential.

The mass matrices M_{u1} and M_{u2} can be written in a more elegant way with

$$\begin{aligned} M_{u1} &= V^\dagger_{CKM}\hat{M}_u - \frac{e^{i\delta}}{\sin\delta} Im(V^\dagger_{CKM})\hat{M}_u \,, \\ M_{u2} &= \frac{1}{\sin\delta} Im(V^\dagger_{CKM})\hat{M}_u \,. \end{aligned} \tag{9}$$

Alternatively, a model can be constructed with two Higgs doublets coupling to the down sector and one Higgs doublet coupling to the up sector.[1] We will concentrate on the above scenario for detailed discussions.

We now go further to construct a realistic model. A common problem for models with spontaneous CP violation is that a strong QCD θ term will be generated.[12] Constraints from neutron dipole moment measurements will rule out spontaneous CP violation as the sole source if there is no mechanism to make sure that the θ term is small enough. The model mentioned above faces the same problem. We therefore supplement the model with a Peccei-Quinn (PQ) symmetry[13] to ensure a small θ.

To have spontaneous CP violation and also PQ symmetry simultaneously, more than two Higgs doublets are needed.[14] For our purpose we find that in order to have spontaneous CP violation with PQ symmetry at least three Higgs doublets $\phi_i = e^{i\theta_i}H_i$ and one complex Higgs singlet $\tilde{S} = e^{i\theta_s}S = e^{i\theta_s}(v_s + R_s + iA_s)/\sqrt{2}$ are required. The Higgs singlet with a large VEV renders the axion from PQ symmetry breaking to be invisible,[15,16] thus satisfying experimental constraints on axion couplings to fermions. We will henceforth work with models with an invisible axion[15] with PQ charges for various fields given by

$$Q_L : 0\,,\ U_R : -1\,,\ D_R : -1\,,\ \phi_{1,2} : +1\,,\ \phi_d = \phi_3 : -1\,,\ \tilde{S} : +2\,.$$

The Higgs potential is

$$
\begin{aligned}
V = & -m_1^2 H_1^\dagger H_1 - m_2^2 H_2^\dagger H_2 - m_3^2 H_3^\dagger H_3 - m_{12}^2 (H_1^\dagger H_2 e^{i(\theta_2-\theta_1)} + h.c.) - m_s^2 S^\dagger S \\
& + \lambda_1 (H_1^\dagger H_1)^2 + \lambda_2 (H_2^\dagger H_2)^2 + \lambda_t (H_3^\dagger H_3)^2 + \lambda_s (S^\dagger S)^2 \\
& + \lambda_3 (H_1^\dagger H_1)(H_2^\dagger H_2) + \lambda_3' (H_1^\dagger H_1)(H_3^\dagger H_3) + \lambda_3'' (H_2^\dagger H_2)(H_3^\dagger H_3) \\
& + \lambda_4 (H_1^\dagger H_2)(H_2^\dagger H_1) + \lambda_4' (H_1^\dagger H_3)(H_3^\dagger H_1) + \lambda_4'' (H_2^\dagger H_3)(H_3^\dagger H_2) \\
& + \frac{1}{2}\lambda_5 ((H_1^\dagger H_2)^2 e^{i2(\theta_2-\theta_1)} + h.c.) + \lambda_6 (H_1^\dagger H_1)(H_1^\dagger H_2 e^{i(\theta_2-\theta_1)} + h.c.) \\
& + \lambda_7 (H_2^\dagger H_2)(H_1^\dagger H_2 e^{i(\theta_2-\theta_1)} + h.c.) + \lambda_8 (H_3^\dagger H_3)(H_1^\dagger H_2 e^{i(\theta_2-\theta_1)} + h.c.) \\
& + f_1 H_1^\dagger H_1 S^\dagger S + f_2 H_2^\dagger H_2 S^\dagger S + f_3 H_3^\dagger H_3 S^\dagger S \\
& + d_{12} (H_1^\dagger H_2 e^{i(\theta_2-\theta_1)} + H_2^\dagger H_1 e^{-i(\theta_2-\theta_1)}) S^\dagger S \\
& + f_{13} (H_1^\dagger H_3 S e^{i(\theta_3+\theta_s-\theta_1)} + h.c.) + f_{23} (H_2^\dagger H_3 S e^{i(\theta_3+\theta_s-\theta_2)} + h.c.) \,. \qquad (10)
\end{aligned}
$$

Note that only two independent phases occur in the above expression, which we choose to be $\delta = \theta_2 - \theta_1$ and $\delta_s = \theta_3 + \theta_s - \theta_2$. The phase $\theta_3 + \theta_s - \theta_1$ can be written as $\delta + \delta_s$. Differentiating with respect to δ_s to get one of the conditions for minimization of the potential, we get

$$
f_{13} v_1 v_3 v_s \sin(\delta_s + \delta) + f_{23} v_2 v_3 v_s \sin \delta_s = 0 \,. \qquad (11)
$$

It is clear that δ and δ_s are related with

$$
\tan \delta_s = -\frac{f_{13} v_1 \sin \delta}{f_{23} v_2 + f_{13} v_1 \cos \delta} \,. \qquad (12)
$$

The phase δ is the only independent phase in the Higgs potential. A non-zero $\sin \delta$ is the source of spontaneous CP violation and also the only source of CP violation in the model.

Removing the would-be Goldstone bosons, one can write the Yukawa interactions for physical Higgs degrees of freedom as the following[1]

$$
\begin{aligned}
L_Y = & \bar{U}_L [\hat{M}_u \frac{v_1}{v_{12} v_2} - (\hat{M}_u - V_{CKM} Im(V_{CKM}^\dagger) \hat{M}_u \frac{e^{i\delta}}{\sin \delta}) \frac{v_{12}}{v_1 v_2}] U_R (H_1^0 + i a_1^0) \\
& + \bar{U}_L \hat{M}_u U_R [\frac{v_3}{v_{12} v}(H_2^0 + i a_2) - \frac{1}{v} H_3^0 + \frac{v_3^2}{v^2 v_s}(H_4^0 + i a)] \\
& - \bar{D}_L \hat{M}_d D_R [\frac{v_{12}}{v_3 v}(H_2^0 - i a_2) + \frac{1}{v} H_3^0 + \frac{v_{12}^2}{v^2 v_s}(H_4^0 - i a)] \\
& + \sqrt{2} \bar{D}_L [V_{CKM}^\dagger \hat{M}_u \frac{v_1}{v_2 v_{12}} - (V_{CKM}^\dagger \hat{M}_u - Im(V_{CKM}^\dagger) \hat{M}_u \frac{e^{i\delta}}{\sin \delta}) \frac{v_{12}}{v_1 v_2}] U_R H_1^- \\
& - \sqrt{2} \frac{v_3}{v_{12} v} \bar{D}_L V_{CKM}^\dagger \hat{M}_u U_R H_2^- - \sqrt{2} \frac{v_{12}}{v v_3} \bar{U}_L V_{CKM} \hat{M}_d D_R H_2^+ + h.c. \,. \qquad (13)
\end{aligned}
$$

Here a is the axion. The fields, H_i^0 and a_i^0 are not the mass eigenstates yet and will be mixed in the potential.

Note that the couplings of a and H_4^0 to quarks are suppressed by $1/v_s$, and that only the exchange of H_1^0 and a_1^0 can induce tree level FCNC interactions. The FCNC coupling is proportional to $V_{CKM} Im(V_{CKM}^\dagger)\hat{M}_u$.

For the model presented here, FCNCs only involve the up quark sector. The most stringent constraint on the Higgs mass comes from $D^0 - \bar{D}^0$. We find that the Higgs mass can be as low as a hundred GeV from this constraint. Such low Higgs mass can be probed at LHC and ILC.

The neutron EDM d_n provides much information on the model parameters. The standard model predicts a very small[17] d_n ($< 10^{-31} e$ cm). The present experimental upper bound on neutron EDM d_n is very tight:[5] $|d_n| < 0.63 \times 10^{-25} e$ cm. In the model considered above, the quark EDMs will be generated at loop levels due to mixing between a_i and H_i.

The one loop contributions to the neutron EDM are suppressed for the usual reason of being proportional to light quarks masses to the third power for diagram in which the internal quark is the same as the external quark. In our model, there is a potentially large contribution when there is a top quark in the loop. However, the couplings to top are proportional to s_{13}, therefore the contribution to neutron EDM is much smaller than the present upper bound. It is well known that exchange of Higgs at the two loop level may be more important than the one loop contribution, through the quark EDM, quark color EDM, and the gluon color EDM. We find that in the model discussed above, the neutron EDM can reach the present experimental bound. Improved measurement on neutron EDM can provide us with more information.

3. Conclusion

To summarize, I have described a model in which the CP violating phase in the CKM mixing matrix to be the same as that causing spontaneous CP violation in the Higgs potential. When the Higgs boson masses are set to be very large, the phase in the CKM matrix can be made finite and CP violating effects will not disappear. An interesting feature of this model is that the FCNC Yukawa couplings are fixed in terms of the quark masses and CKM mixing angles, making phenomenological analysis much easier.

Acknowledgments

This work was supported in part by NSC and NCTS. I thank S.-L. Chen, N.G. Deshpande, J. Jiang and L.-H. Tsai for collaboration on the work reported here.

References

1. Sho-Long Chen, N.G. Deshpande, Xiao-Gang He, Jing Jiang, Lu-Xing Tsai, arXiv:0705.1061[hep-ph].

2. T. D. Lee and C.-N. Yang, Phys. Rev. **104**, 254(1956).
3. S. L Glashow, Nucl. Phys. **22**, 579(1967); S. Weinberg, Phys. Rev. Lett. **19**, 1264(1967); A. Salam, p. 367 of Elementary Particle Thoery, ed. N. Svartholm (Almquist and Wiksells, Stockholm, 1969).
4. J. H. Christenson et al., Phys. Rev. Lett. **13**, 138(1964).
5. W-M Yao *et al* 2006 J. Phys. G: Nucl. Part. Phys. **33** 1.
6. M. Kobayashi and T. Maskawa, Prog. Theor. Phys. **49**, 652 (1973); N. Cabbibo, Phys. Rev. Lett. **10**, 531(1963).
7. L. Wolfenstein, Pphys. Rev. Lett. 51, 1945(1983).
8. M. Bona et al., http://utfit.romal.infn.it
9. T. D. Lee, Phys. Rev. D **8**, 1226 (1973); T. D. Lee, Phys. Rept. **9**, 143 (1974).
10. S. Weinberg, Phys. Rev. Lett. **37**, 657 (1976); G. C. Branco, Phys. Rev. Lett. **44**, 504 (1980).
11. D. Chang, X. G. He and B. H. J. McKellar, Phys. Rev. D **63**, 096005 (2001) [arXiv:hep-ph/9909357]; G. Beall and N. G. Deshpande, Phys. Lett. B **132**, 427 (1983); I. I. Y. Bigi and A. I. Sanda, Phys. Rev. Lett. **58**, 1604 (1987).
12. R. Akhoury and I. I. Y. Bigi, Nucl. Phys. B **234**, 459 (1984).
13. R. D. Peccei and H. R. Quinn, Phys. Rev. D **16**, 1791 (1977): R. D. Peccei and H. R. Quinn, Phys. Rev. Lett. **38**, 1440 (1977).
14. X. G. He and R. R. Volkas, Phys. Lett. B **208**, 261 (1988) [Erratum-ibid. B **218**, 508 (1989)]; C. Q. Geng, X. D. Jiang and J. N. Ng, Phys. Rev. D **38**, 1628 (1988).
15. A.R. Zhitnitsky, Sov. J. Nucl. Phys. **31**, 260(1980); M. Dine, W. Fischler and M. Srednicki, Phys. Lett. B **104**, 199 (1981).
16. J. E. Kim, Phys. Rev. Lett. **43** (1979) 103; M. Shifman, A. Vainshtein, V. Zakharov, Nucl. Phys. **B166** (1980) 493.
17. X. G. He, B. H. J. McKellar and S. Pakvasa, Int. J. Mod. Phys. A **4**, 5011 (1989) [Erratum-ibid. A **6**, 1063 (1991)]; B. H. J. McKellar, S. R. Choudhury, X. G. He and S. Pakvasa, Phys. Lett. B **197**, 556 (1987).

QUANTUM COMPUTATION IN SILICON — DEVICE MODELING, TRANSPORT AND FAULT-TOLERANCE

L. C. L. HOLLENBERG*

Centre for Quantum Computer Technology, School of Physics, University of Melbourne, Parkville, Victoria 3010, Australia
**E-mail: lloydch@unimelb.edu.au*

The Kane concept of quantum computation using single phosphorus donors in silicon has revolutionized the way we think about P-doped silicon devices. However, because of the enormous effort required in fabricating such devices using single atom techniques, detailed modeling and assessment of fault-tolerant operation is required to determine and optimize the scalability, and indeed feasibility, of such a vision.

Keywords: Quantum computing, donors in silicon, nanoelectronics.

1. Introduction

The scale-up of an array of qubits to a fully fledged quantum computer operating fault-tolerantly necessarily assumes concatenation of encoding and quantum error correction, and is exceedingly complex. A schematic showing the recursive requirements of concatenation from physical level to logical level quantum error correction is shown in Figure 1. Any qubit proposal must consider the ultimate physical implementation of these requirements before claiming scalability. In this paper we review the proposal for quantum computing in silicon which embraces scalability in this strong sense, and some of the theoretical underpinnings of these concepts.

2. Donor Based Qubits in Silicon

The original Kane concept[1] for quantum computing based on the control of single donor qubits in silicon (Figure 2), was a landmark piece of work which spawned a number of proposals for encoding and manipulating quantum information based on donor spin[2,3] or charge degrees of freedom[4] have been put forward. Recently, these developments have culminated in a 2D donor based architecture[5] (shown in Figure 3), which, by incorporating transport, may be the first truly scalable design in silicon. Because of the significant effort required in single atom fabrication[6–9] and quantum measurement to bring such a vision to reality, a number of significant theoretical challenges must be met to critically assess the scalability of these

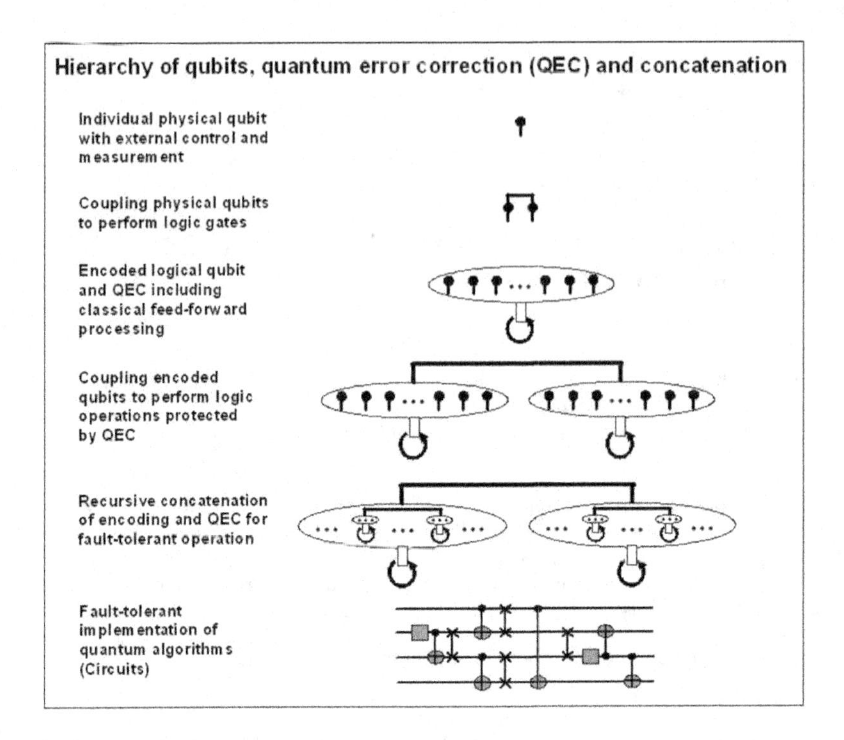

Fig. 1. Schematic showing the recursive application of quantum error correction in fault-tolerant quantum computer design.

architectures and describe emerging small scale experimental devices. In particular microscopic modelling of the basic quantum control of donor electrons through wave function engineering plays a major role in determining the essential parameters, such as: gate times, cross-talk, gate fidelity and effects of decoherence.

3. Donor Qubit Spin System

The effective spin level description of the donor qubit system, from which we are able to construct quantum logic gates, can be written down as:

$$\begin{aligned} H &= \mu_B B_z(\sigma^z_{e_1} + \sigma^z_{e_2}) - g_n \mu_n B_z(\sigma^z_{n_1} + \sigma^z_{n_2}) \\ &\quad + A_1(t)\vec{\sigma}_{e_1}\cdot\vec{\sigma}_{n_1} + A_2(t)\vec{\sigma}_{e_2}\cdot\vec{\sigma}_{n_2} + J_{12}(t)\vec{\sigma}_{e_1}\cdot\vec{\sigma}_{e_1} + H_{\mathrm{ac(t)}} \end{aligned} \tag{1}$$

where the control functions $A(t)$ and $J(t)$ are the hyperfine and exchange couplings controlled by the A and J gates respectively, and $H_{ac}(t)$ is the RF Hamiltonian which flips nuclear spins when they are brought into resonance via the A-gates. From this Hamiltonian one can carry out NMR type analyses of single and two

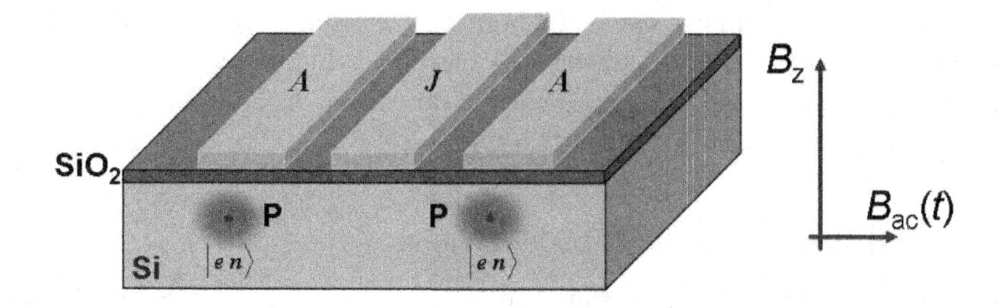

Fig. 2. The basic two-qubit system of the Kane proposal. Qubit states are encoded on the nuclear spins of donor P atoms in silicon, while single qubit gates are carried out via the A-gate which Stark shift the hyperfine interaction and bring the qubit into resonance with a RF field $B_{ac}(t)$. Two qubit gates are carried out via J-gate deformation of the electron wave function which changes the exchange interaction and mediates entangling gates.

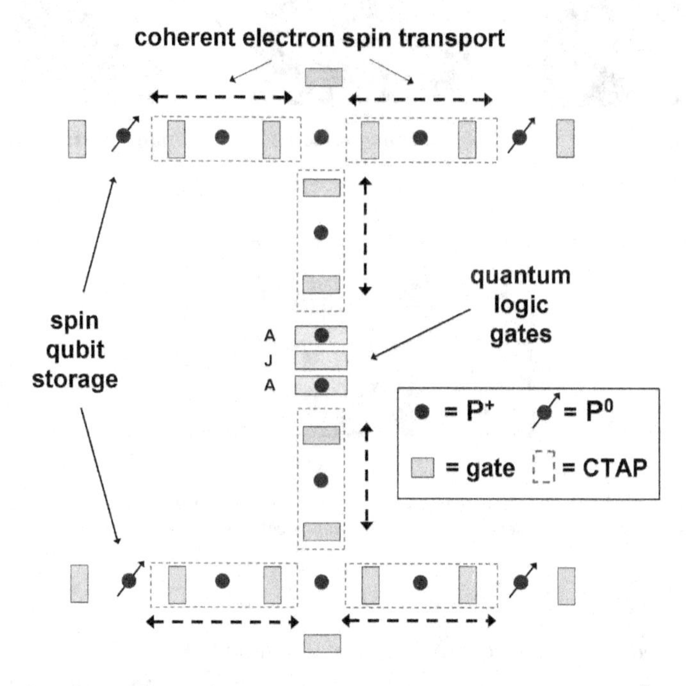

Fig. 3. The two dimensional donor electron spin quantum computer architecture[5] incorporating coherent transport by adiabatic passage (CTAP) along ionized P+ pathways. Distinct storage and interaction regions (A-J-A) give non-local qubit interactions for scale-up.

qubit gates assuming continuous control of the functions $A(t)$ and $J(t)$. These functions represent precise control of the donor electron wave function via surface gate potentials. The nae expectation is that by literally guiding the donor electron wave function away from the nucleus using the A-gate one reduces the hyperfine interaction strength, or by bringing two neighbouring electron wave functions closer together the exchange interaction strength is increased. In reality, understanding the extent to which qubit control can be carried out using surface gates requires a detailed microscopic treatment of the entire system.

4. Device Modeling – General Considerations

In order to understand qubit control from the physical level we must describe in detail how the donor electron wave function can be manipulated by an inhomogeneous gate field to produce the required quantum gates. A schematic showing the various aspects of the general controlled donor modeling problem is shown in Figure 4.

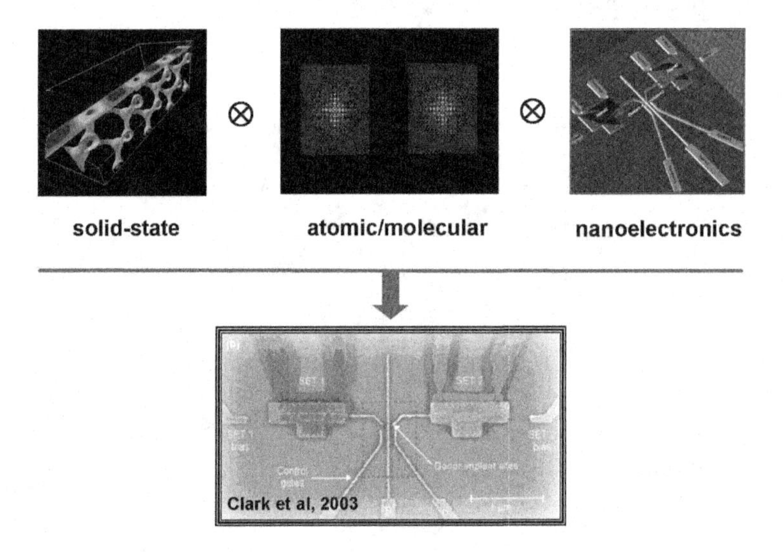

Fig. 4. Modelling of donor qubit devices in general requires the combined treatment of the underlying solid-state silicon environment, the molecular few electron P-P system, and the nanoelectronic control circuit.

In total, one requires three components in a unified description: the underlying solid-state description of the silicon crystal, the molecular states of the P-P system and the gate fields generated by the surface nano-electronic circuit. The various calculations reported after the Kane proposal was published typically neglected one or more of these essential components. Initially, the hydrogenic description, as employed by Kane in the original proposal, was used to describe the effect of surface gates on the donor electron wave function. The importance of including the silicon Bloch structure in the donor electron wave function was demonstrated dramatically when the un-gated exchange coupling was computed and shown to have relatively large oscillations as a result of small atomic scale deviations from an "ideal" placement.[10,11] We turn to this problem in detail.

5. Exchange Oscillations

Since the donor electron exists in a crystalline environment, only the envelope of the wave function is hydrogenic with an effective Bohr radius of approximately 2nm. The wave function itself oscillates on a scale that is small compared to the silicon lattice constant. The result is that the exchange interaction strength based on an overlap of the two donor electron wave functions is not only dependent on the separation of the donors, but highly dependent on the actual lattice sites at which the localising P donors reside. This sensitivity to placement at the atomic site level, first pointed out in ref,[10] implies that there will never be uniformity of the exchange interaction. This is not a critical problem as such for the quantum computer, since characterisation methods[12–14] will be required to precisely construct quantum logic gates in any case. However we do need to understand the phenomenon and know that it is not an artefact of an over-simplified treatment of the P-P system.

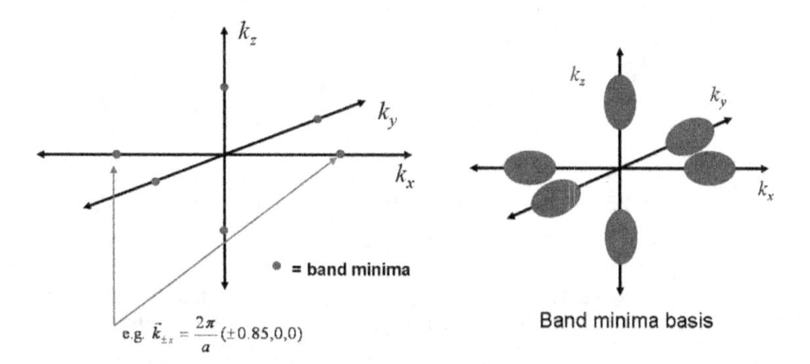

Fig. 5. On the left the conduction the Bloch states at the six degenerate band minima points in k-space are taken as the description of the donor electron in the effective mass approximation. In the band minima basis treatment (right) a large basis of Bloch states around each minimum is used.

The original calculation in Ref. 10 used an effective mass treatment of the donor electron system, in conjunction with an approximate form of the Heitler-London (HL) treatment, in the absence of gate fields. In the effective mass approximation the donor electron wave function is written as a superposition over the Bloch states at the six degenerate conduction band minima (Figure 5 left):

$$\psi(\vec{r}) = \frac{1}{\sqrt{6}} \sum_{\mu=1}^{6} F_\mu(\vec{r}) e^{i\vec{k}_\mu \cdot \vec{r}} u_{\vec{k}_\mu}(\vec{r}), \tag{2}$$

where the components of the Bloch states with periodicity of the lattice are

$$u_{\vec{k}_\mu}(\vec{r}) = \sum_{\vec{G}} A_{\vec{k}_\mu}(\vec{G}) e^{i\vec{G}\cdot\vec{r}}. \tag{3}$$

As a first approximation, the band minima momenta components dominate the long distance behaviour of the state and the oscillatory Bloch components can be ignored. In Ref. 11 the first ab-initio treatment of the exchange oscillations was carried out by including Bloch states computed in the pseudopotential framework, and using the full HL integral. The results (Figure 6) did not vary much from Ref. 10, but verified the exchange oscillations effect to the next level of approximation.

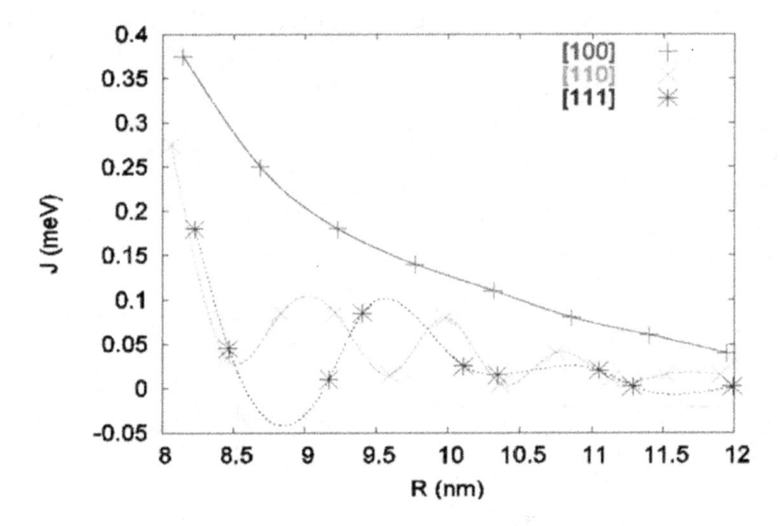

Fig. 6. Effective mass results (Ref. 11) for the exchange coupling strength as a function of donor separation in the [100], [110] and [111] directions. Data points indicate lattice sites.

The first calculation beyond the effective mass approximation was carried out in Ref. 15 using a framework which includes a large basis of states around each of the conduction band minima (Figure 5 right). Although solving the Coulomb problem in momentum space has its challenges (one must regulate the singularity correctly), the advantage is that the donor core correction and surface gate fields can be added at the Hamiltonian level and the system then diagonalised.

The results for the exchange coupling along the direction [110], which displays large exchange oscillations in the effective mass approximation, are given in Figure 7. One sees immediately that the effect of the inclusion of higher order Bloch states in the wave function is to smooth out the oscillations slightly compared to the effective mass case. The core correction (which correctly reproduces the 1S manifold valley splitting) has a further dampening effect. These results still employ the HL treatment of the P-P system - a complete treatment using a configuration interaction

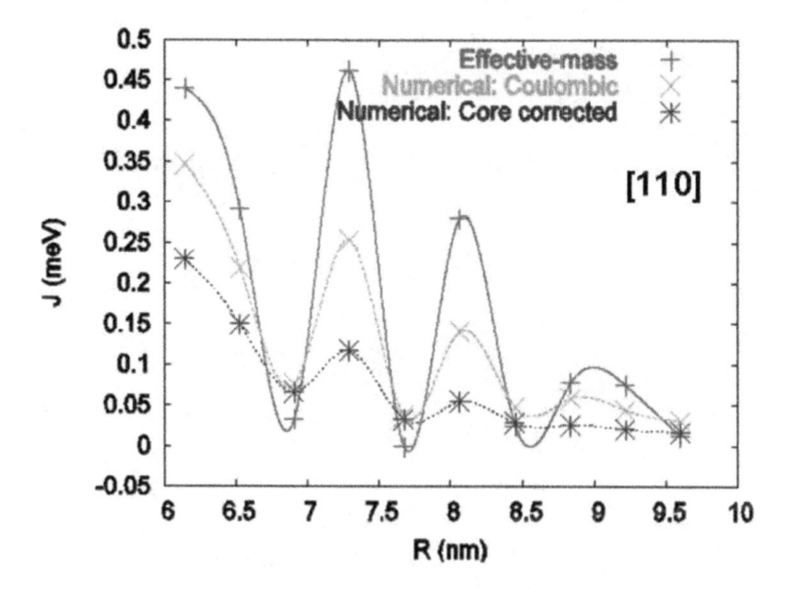

Fig. 7. Band basis results (Ref. [7]) for the exchange coupling along the [110] direction, including the core correction.

calculation is still required to fully understand the extent of the exchange oscillations in realistic, gated devices.

6. Microwave Driven P-P$^+$ System

In terms of fabrication, an important stepping stone to the realization of the donor devices described is the charge qubit based on a singularly ionized P-P+ system.[4] SET readout for such a device is more straightforward than for spin since only charge detection is required (as opposed to a spin-charge transduction step). The first experiments currently underway on the P-P+ system involve microwave spectroscopy. In order to interpret such measurements we must understand the driven P-P+ system at the microscopic level. In analogy with the exchange oscillations, it is perhaps not surprising that this system is also sensitive to atomic level variations in the donor placement.[16] A comprehensive treatment of the driven problem in the presence of charge fluctuation noise sources has been carried out in the effective mass approximation as a starting point.[17] Figure 8 shows the results for the probability of finding the electron on the left donor site in the P-P+ system as the DC field is swept out. Peaks occur when the system is in resonance with the driving microwave AC field. The different curves show how the results will vary as the donor placement is varied by only a few lattice sites. These calculations are now being extended to include a full gate structure simulation[18] and higher level treatment of the molecular P-P+ states.

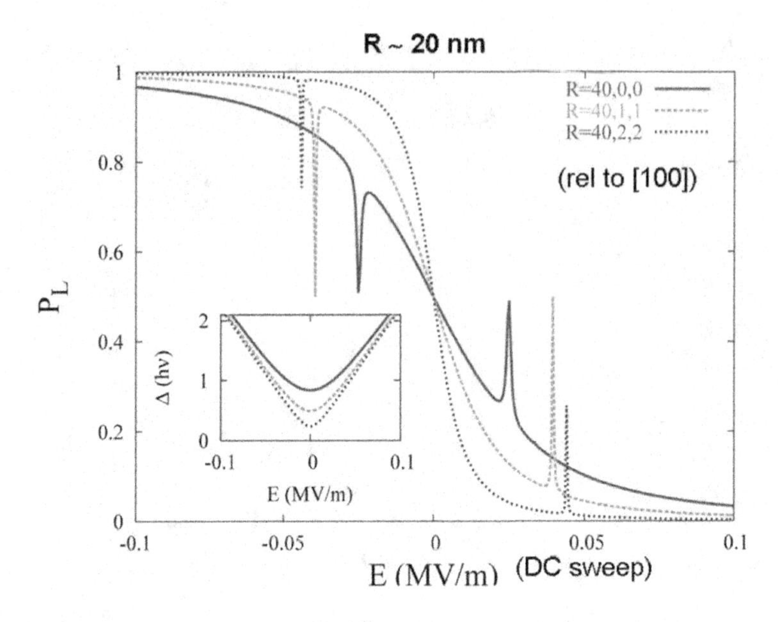

Fig. 8. Calculation of the probability of finding the electron of the P-P+ system at the left donor as a function of the DC field strength (and direction) in the presence of an AC microwave driving field at 40GHz.

There are developing frameworks in which to bring all these aspects of the problem into one unified description - e.g. within the band-minima basis expansion,[15] or based on an optimised tight-binding device simulation.[19] Such tools will be invaluable in analysing the complex nature of quantum control, error correction cross-talk and scale-up in these devices. In the meantime, and under certain approximations, a first unifying treatment with detailed control structure simulations has been completed for the microwave driven P-P+ charge qubit device[18] to the level that detailed comparisons of coherent electron control with experiment can be made. As experiments come closer to realizing the Kane devices, it is critical that increasingly sophisticated modelling techniques are brought to bear on the gated donor systems. This in turn will drive up the level of realism that can be included in fault-tolerant quantum computer architecture design, and ultimately a blueprint for a working device.

7. Coherent Transport in Donor Systems

In Ref. 5 a mechanism for donor electron spin transport was proposed based on coherent charge transport by adiabatic passage (CTAP),[20] which we briefly review here. Consider an ionized chain of three donors with one electron, with controllable tunnel couplings between the donors, and energies of the donor states, as per Figure 9. This control would be effected by surface gates.

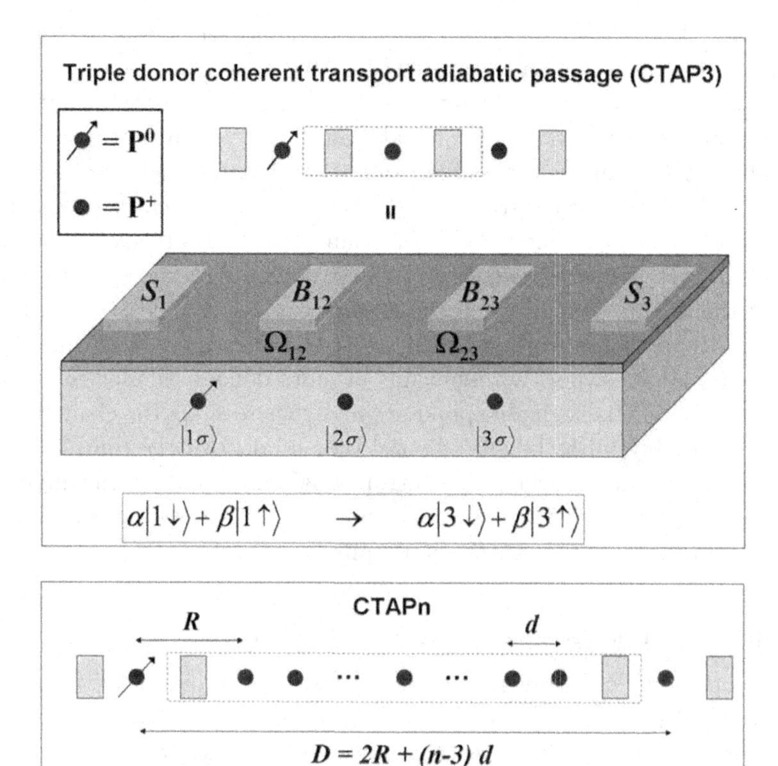

Fig. 9. Coherent transport by adiabatic passage (CTAP), defined with respect to P donors in Si. Ionized donors provide a well defined pathway for electron transport from donor 1 to 3 without populating the intervening donor.

The effective Hamiltonian for this system can be written as:

$$H = \Delta|2\rangle\langle 2| - \hbar\,(\Omega_{12}|1\rangle\langle 2| + \Omega_{23}|2\rangle\langle 3| + \text{h.c.})\,, \tag{4}$$

where $\Omega_{ij} = \Omega_{ij}(t)$ is the coherent tunneling rate between donors i and j and $\Delta = E_2 - E_1 = E_2 - E_3$. This system can be solved analytically, and comprises three eigenstates:

$$\begin{aligned}
|D_+\rangle &= \sin\Theta_1 \sin\Theta_2|1\rangle + \cos\Theta_2|2\rangle + \cos\Theta_1 \sin\Theta_2|3\rangle,\\
|D_-\rangle &= \sin\Theta_1 \cos\Theta_2|1\rangle - \sin\Theta_2|2\rangle + \cos\Theta_1 \cos\Theta_2|3\rangle,\\
|D_0\rangle &= \cos\Theta_1|1\rangle - \sin\Theta_1|3\rangle,
\end{aligned} \tag{5}$$

with energies $\epsilon_\pm$ and ϵ_0 accordingly. The mixing angles in these expressions are set by the tunneling rates between localised states. The separations between these eigenstates are set by the tunneling matrix elements. The state $|D_0\rangle$ is a superposition of the particle being in the left and right donors, and can be varied smoothly between these two sites as long as the adiabatic condition is satisfied. This allows coherent transport across the three-donor system without ever populating the central donor. By virtue of its adiabatic nature the CTAP protocol is robust and is in fact relatively quick - theoretical calculations based on the microscopic analysis of gate assisted tunneling between donors indicate high fidelity transport over typical distances of about 100nm in 2ns is possible.[5] In a real system, transport fidelity will be controlled by the adiabatic criterion and the timescale of any charge dephasing, with respect to the controlled inter-donor tunnelling rates. After solving the time dependence of the system in the presence of a charge dephasing term, we found quite generally that the fidelity is high when the transport time is at least an order of magnitude faster than charge dephasing. Higher order versions of CTAP have been developed which allow transport across chains of more donors. Similar robustness is found, and again, there is negligible population anywhere along the chain, except at the ends. Remarkably, with the exception of the end donors, the central donors in the chain do not require surface gate control, provided that the tunneling rates are high. It is this feature which enables the overall gate density of the computer to be reduced by incorporating CTAP transport rails.

8. Quantum Computer Architectures and Beyond

Applying this scheme to spin transport opens a new vista of opportunities for designing 2D architectures for donor based quantum computing (see Figure 3). In particular, all of the scalability issues with donor based proposals can be addressed, and the realm of fault-tolerant donor based architectures explored.[5] CTAP donor chains provide flexibility in reducing the overall gate density, and valuable space for the large SET readout and nanoelectronic control structures. Transport rails also allow for non-local qubit-qubit interactions, and the possibility of bypassing non-functioning interaction regions for defect-tolerant computing. Finding the optimal geometrical arrangement of qubits in conjunction with transport requires a high level analysis of fault-tolerant operation, including concatenation of logical qubit encoding.

A promising arrangement shown in Figure 10 is the bi-linear array, upon which we are able to compile quantum error correction protocols incorporating physical qubit transport and determine the failure rate for each level of concatenation in an extended rectangle analysis. The result of such calculations are lower bounds for the fault-tolerant error thresholds, to be reported in a forthcoming publication.

Acknowledgments

I would like express my sincere thanks to Prof. Bruce McKellar for his unstinting support, mentorship and friendship, and for the outstanding scholarly example he has set over the past quarter decade. Thanks too to A/Prof. Girish Joshi for his support, friendship and many a nice conversation over the years, which would often make my day. This work was supported by the Australian Research Council, the

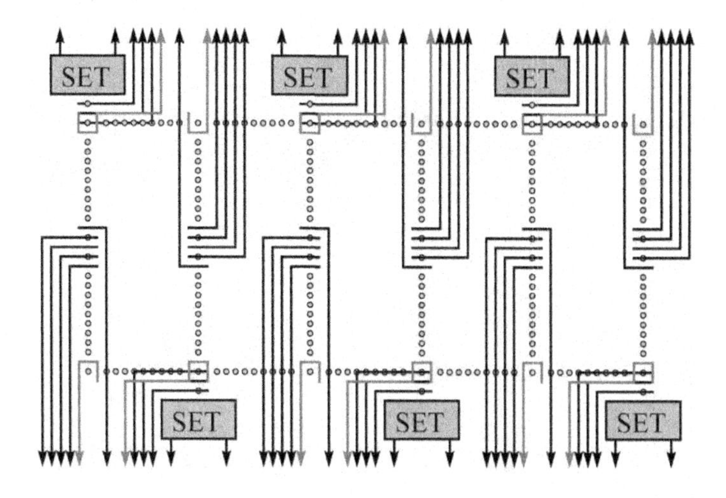

Fig. 10. The bi-linear donor archiecture for fault-tolerant quantum computation in silicon.

Australian Government, and the US National Security Agency (NSA), Advanced Research and Development Activity (ARDA), and the Army Research Office (ARO) under contract number W911NF-04-1-0290.

References

1. B. E. Kane, *Nature* **393**, p. 133 (1998).
2. R. Vrijen et al, *Phys. Rev. A* **62**, p. 012306 (2000).
3. R. de Sousa, J. Delgado and S. Das Sarma, *Phys. Rev. A* **70**, p. 052304 (2004).
4. L. Hollenberg et al, *Phys. Rev. B* **69**, p. 113301 (2004).
5. L. Hollenberg et al, *Phys. Rev. B* **74**, p. 045311 (2006).
6. J. Tucker et al, *Solid State Electronics* **42**, p. 1061 (1998).
7. S. Schofield et al, *Phys. Rev. Lett.* **91**, p. 136104 (2003).
8. F. Ruess et al, *Nano Letters* **4**, p. 1969 (2004).
9. D. Jamieson et al, *Appl. Phys. Lett.* **86**, p. 202101 (2005).
10. B. Koiller, X. Hu and S. Das Sarma, *Phys. Rev. Lett.* **99**, p. 027903 (2002).
11. C. Wellard et al, *Phys. Rev. B* **68**, p. 195209 (2003).
12. J. Cole, S. Devitt and L. Hollenberg, *J. Phys. A: Math. Gen.* **39**, p. 14649 (2006).
13. J. Cole et al, *Phys. Rev. A* **73**, p. 062333 (2006).
14. S. Devitt, J. Cole and L. Hollenberg, *Phys. Rev. A* **73**, p. 052317 (2006).
15. C. Wellard and L. Hollenberg, *Phys. Rev. B* **72**, p. 085202 (2005).
16. X. Hu, B. Koiller and S. Das Sarma, *Phys. Rev. B* **71**, p. 235332 (2005).
17. C. Wellard, L. Hollenberg and S. Das Sarma, *Phys. Rev. B* **74**, p. 075306 (2006).
18. V. Conrad, PhD thesis, University of Melbourne, (Melbourne, Australia, 2007).
19. A. Martins et al, *Phys. Rev. B* **72**, p. 193204 (2005).
20. A. Greentree et al, *Phys. Rev. B* **70**, p. 235317 (2004).

BORN RECIPROCITY IN TWO DIMENSIONS: QUAPLECTIC GROUP SYMMETRY, BRANCHING RULES AND STATE LABELLING

P. D. JARVIS*

School of Mathematics and Physics, University of Tasmania,
Hobart Tas 7001, Australia
E-mail: Peter.Jarvis@utas.edu.au
**Alexander von Humboldt Fellow*

We consider the continuous symmetry group underlying Born's reciprocity principle, namely the so-called quaplectic group, the semidirect product of time-space-energy-momentum coordinate transformations with the Weyl-Heisenberg group. In two dimensional Minkowski space this group is $Q(1,1) \cong U(1,1) \ltimes H(2)$, or in Euclidean space $Q(2) \cong U(2) \ltimes H(2)$. For the 'scalar' system in the sense of induced representations, unitary irreducible representations are carried on a Fock space equivalent to that used by Schwinger as a model of the $SU(2)$ angular momentum algebra, or by Holman and Biedenharn as a model of $SU(1,1)$. Using this construction we consider the branching rules and state labelling problem for the reduction of $Q(2)$ and $Q(1,1)$ to the 'physical' Euclidean and Poincaré subalgebras, respectively. The results serve to illustrate the difficulties of any consideration of Born reciprocity as an extended symmetry principle of nature.

Keywords: Born reciprocity; Heisenberg algebra; two-dimensional oscillator; induced representations.

1. Introduction

Born reciprocity[1] is based on the observation of the apparent exchangeability of "position" and "momentum" in much of the formalism of classical and quantum physics, and seeks to elevate this equivalence to a fundamental principle. The idea of Born[1,2] and Green[2,3] was to formalise this by extending the Minkowski metric of Einstein's special relativity to an invariant metric on "phase space coordinates"

$$d\ell^2 = ds^2 + \frac{c^4}{b^2}dm^2 = dx^\mu dx_\mu + \frac{c^2}{b^2}dp^\mu dp_\mu, \tag{1}$$

where $\quad x^\mu = (ct, \vec{x}), \quad p^\mu = \left(\frac{E}{c}, \vec{p}\right), \quad \eta_{\mu\nu} = \text{diag}\,(+,-,-,\cdots), \quad \mu = 0,1,2,\cdots,$

which can be seen as introducing a new fundamental constant, here a maximal universal unit of force $b > 0$ (which can also be thought of in terms of fundamental constants of acceleration, or length, or time, depending on the interpretation). Born and Green sought reciprocally invariant "master" equations whose zeroes were interpreted via multi-mass relativistic wave equations for the meson spectrum.

What happened to reciprocity? Born and Green's intense work in Edinburgh (together with Fuchs[5]) in the late 'forties and early 'fifties resulted in several papers, after which they appeared to abandon the subject. Possibly, their interest was superseded on the theoretical side by the rise of quantum electrodynamics, together with better theoretical understanding of relativistic wave equations, initiated by the work of Wigner.[6] On the experimental side, the plethora of new 'mesons' made it unlikely that any of these could be regarded as truly 'elementary' particles as their theory required. Not the least important factor was no doubt Green's departure for Adelaide.[7] However, Green returned to the subject in the mid 'sixties with a presentation on relativistic oscillators at the Kyoto high energy meeting,[4] and subsequently Bracken and Green worked on a trilocal model for baryons, where the substituents obeyed reciprocity.[8] Born, shortly before his death in 1970, was apparently supportive of this[a], and indeed his convictions appear engraved on his tombstone in Göttingen[5] under the motto '$pq - qp = h/2\pi i$'.

Born-Green reciprocity can be viewed[9–11] as an alternative paradigm for generalised wave equations, which specify unitary irreducible representations of the full symmetry group, in the same way that relativistic wave equations establish unitary irreducible representations of the Poincaré group in four dimensions.[6] It can be argued[9–11] that the appropriate invariance group is the so-called quaplectic group $Q(3,1) \cong U(3,1) \ltimes H(4)$, or more generally in D spacetime dimensions, the group $Q(D-1,1) \cong U(D-1,1) \ltimes H(D)$ of reciprocal relativity, the semi-direct product of the pseudo-unitary group of linear transformations between x^μ and p^μ which preserve both the extended metric $d\ell^2$ and the symplectic form, with the Weyl-Heisenberg group. The Wigner-Mackey method of induced representations can be applied for this case, and $b \to \infty$ contraction limits of the appropriate generalised reciprocally invariant wave equations should collapse to the standard relativistic wave equations of particle physics, as, for example, solutions of the massive Klein-Gordon equation can be seen as going over to the Galilean invariant nonrelativistic wave functions in the $c \to \infty$ limit.[13]

In this note we consider aspects of reciprocity in $1+1$ dimensional spacetime. After a brief discussion of reciprocal-relativistic kinematics to motivate the introduction of the full quaplectic symmetry group of reciprocal relativity, we point out the equivalence between the general Mackey induced representation method, and the physically more intuitive construction with spaces of 'relativistic oscillators'. In two dimensions, the underlying Hilbert space is precisely that used by Schwinger[14] as a model of angular momentum and the $SU(2)$ algebra, or alternatively by Holman and Biedenharn[15] as a model of the $SU(1,1)$ algebra. We

[a]Expressed in correspondence with Green (Tony Bracken, private communication).

show the connection of these two realisations (on the same space) with the 'scalar' case of unitary irreducible representations of the quaplectic symmetry algebras $Q(2) \cong U(2) \ltimes H(2)$, and $Q(1,1) \cong U(1,1) \ltimes H(2)$ respectively. We discuss aspects of the group branching rules, and operator diagonalisation and basis choices, for the cases $Q(2) \supset E(2) \cong SO(2) \ltimes T(2)$, $Q(1,1) \supset P(1,1) \cong SO(1,1) \ltimes T(2)$ respectively, which would be involved in any complete theory of reciprocal relativity viewed as a generalised symmetry principle for 'elementary particles'. More general remarks about the further physics implications of the reciprocity principle are given in the conclusions.

2. Two Dimensional Reciprocal Relativity

The generalisation of symmetry transformations to include 'momentum-space' and 'energy-time' operations, along the lines suggested by (1), obviously implies that the description entails the kinematics of non-inertial, interacting systems (presumably with *non-gravitational* forces, if any local description is to avoid the exigencies of the equivalence principle). In one space and one time dimension, standard non-relativistic force, power, and velocity relations between moving and interacting systems can be written in terms of transformations of time/space and momentum/energy differentials[11] as

$$\begin{pmatrix} dt \\ dq \\ dp \\ d\varepsilon \end{pmatrix} \to \begin{pmatrix} 1 & 0 & 0 & 0 \\ -v & 1 & 0 & 0 \\ -f & 0 & 1 & 0 \\ r & f & -v & 1 \end{pmatrix} \begin{pmatrix} dt \\ dq \\ dp \\ d\varepsilon \end{pmatrix}$$

whereby the line element dt^2, and the antisymmetric bilinear form $\omega = dt \wedge d\varepsilon - dq \wedge dp$, are preserved. Under the special-relativistic generalisation

$$\begin{pmatrix} dt \\ dq \\ dp \\ d\varepsilon \end{pmatrix} \to \frac{1}{\sqrt{1-v^2/c^2}} \begin{pmatrix} 1 & -\frac{v}{c^2} & 0 & 0 \\ -v & 1 & 0 & 0 \\ 0 & 0 & 1 & -\frac{v}{c^2} \\ 0 & 0 & -v & 1 \end{pmatrix} \begin{pmatrix} dt \\ dq \\ dp \\ d\varepsilon \end{pmatrix}$$

on the one hand, and the reciprocal-relativistic generalisation

$$\begin{pmatrix} dt \\ dq \\ dp \\ d\varepsilon \end{pmatrix} \to \frac{1}{\sqrt{1-f^2/b^2}} \begin{pmatrix} 1 & 0 & -\frac{f}{b^2} & 0 \\ 0 & 1 & 0 & \frac{f}{b^2} \\ -f & 0 & 1 & 0 \\ 0 & f & 0 & 1 \end{pmatrix} \begin{pmatrix} dt \\ dq \\ dp \\ d\varepsilon \end{pmatrix}$$

on the other, the invariant symmetric metric is extended in the natural way to

$$d\ell^2 = \left(dt^2 - \frac{dq^2}{c^2}\right) + \frac{1}{b^2}\left(\frac{d\varepsilon^2}{c^2} - dp^2\right)$$

involving new constants c, b denoting maximum relative coordinate speed, and maximum relative force (momentum exchange rate), respectively. However, both of these forms still preserve the (symplectic) antisymmetric 2-form ω, meaning that the linear transformations are canonical transformations in 'phase space'. Adopting these invariances as fundamental principles, leads then to the extended homogeneous symmetry group

$$U(1,1) \cong Sp(4,\mathbb{R}) \cap SO(2,2)$$

(which in addition to the above transformations parametrised by f,v and r, admits an additional one-parameter subgroup, a substitutional operation between time/space and energy/momentum, which commutes with Lorentz transformations[11]). Together with global translations on 'phase space', which in the quantum case we might expect to develop a central extension, this leads to the full quaplectic group of reciprocal relativity,

$$Q(1,1) \cong U(1,1) \ltimes H(2),$$

the semidirect product of the $U(1,1)$ homogeneous 'complex Lorentz' group with the two-dimensional Heisenberg-Weyl group (whose Lie algebra is the Heisenberg algebra). For completeness we give here the nine-dimensional Lie algebra of this group. The generators[b] are $\{E^\mu{}_\nu, \overline{Z}^\mu, Z_\nu, \mathsf{I}\}$, $\mu, \nu = 0, 1$:

$$\begin{aligned} [E^\kappa{}_\lambda, E^\mu{}_\nu] &= \delta^\mu{}_\lambda E^\kappa{}_\nu - \delta^\kappa{}_\nu E^\mu{}_\lambda, \\ [E^\mu{}_\nu, \overline{Z}^\kappa] &= \delta^\kappa{}_\nu \overline{Z}^\mu, \qquad [E^\mu{}_\nu, Z_\lambda] = -\delta^\mu{}_\lambda Z_\nu, \\ [Z_\mu, \overline{Z}^\nu] &= \delta_\mu{}^\nu \mathsf{I}, \qquad [Z_\mu, Z_\nu] = 0 = [\overline{Z}^\mu, \overline{Z}^\nu] \end{aligned} \tag{2}$$

with, in unitary representations, $E^\mu{}_\nu{}^\dagger = \eta^{\mu\alpha}\eta_{\nu\beta}E^\beta{}_\alpha$, and $Z_\mu{}^\dagger = \eta_{\mu\alpha}\overline{Z}^\alpha$. The structure is more explicitly expressed in terms of real and imaginary parts,

$$\begin{array}{r} \text{Lorentz algebra} \\ \text{reciprocal boosts} \\ \text{Heisenberg algebra} \end{array} \begin{cases} L_{\mu\nu} = i(E_{\mu\nu} - E_{\nu\mu}); \\ M_{\mu\nu} = \frac{1}{2}(E_{\mu\nu} + E_{\nu\mu}); \\ Z_\mu = \frac{1}{\sqrt{2}}(X_\mu + iP_\mu),\ \overline{Z}_\mu = \frac{1}{\sqrt{2}}(X_\mu - iP_\mu), \end{cases} \tag{3}$$

where of course X^μ and P_ν are the relativistic 'coordinate' and momentum operators respectively, and I is the central extension. The Euclidean counterpart $Q(2) \cong U(2) \ltimes H(2)$ has an analogous algebra, except that indices are conventionally labelled $a, b = 1, 2$ with generators $E^a{}_b$, Z_a, $\overline{Z}^b$, and the commutation relations and hermiticity conditions involve the Euclidean metric δ_{ab}, the ordinary Kronecker δ.

[b] The definitions are the same in $(D-1, 1)$ dimensions with the appropriate extension of the range of subscripts.

The Mackey theory of induced representations in the quaplectic case[9] follows the standard procedure[19] starting with unitary irreducible representations of the normal subgroup, in this case the Weyl-Heisenberg group. The latter (for the central term $\mathsf{I} \to \hbar\mathbb{1}$, $\hbar \neq 0$), are of course well known. Remarkably,[9] the entire Mackey construction boils down to the statement that the general unitary irreducible representations are equivalent to tensor product spaces of the Weyl-Heisenberg group, together with an additional '$U(1,1)$-valued spin' (or simply '$U(2)$-valued' spin in the Euclidean case). Specifically, this 'spin-orbit' type construction can be written for the generators[c] as

$$E^{\mu}{}_{\nu} = Z^{\mu}{}_{\nu} + \mathbb{E}^{\mu}{}_{\nu}, \quad \text{where} \quad Z^{\mu}{}_{\nu} = \frac{1}{2\hbar}\{\overline{Z}^{\mu}, Z_{\nu}\}.$$

Thus $\mathbb{E}^{\mu}{}_{\nu}$ generates $U(1,1)$, but commutes with $Z_{\mu}, \overline{Z}^{\nu}$. We also have the quaplectic Casimir operators $\mathrm{tr}(\mathbb{E}^n)$, for example defining $U := \mathrm{tr}(E) \equiv E^{\mu}{}_{\mu}$,

$$C_1 = \mathrm{tr}(\mathbb{E}) = \mathbb{E}^{\mu}{}_{\mu} = U - \frac{1}{2\hbar}(P{\cdot}P + X \cdot X).$$

3. Quaplectic Group Branching and State Labelling in Two Dimensions

Given the quaplectic group as a generalisation of space-time symmetries, we wish to explore in tutorial detail here the implications of the associated group reduction and state labelling problem in the two-dimensional cases, $Q(1,1) \supset P(1,1) \cong SO(1,1) \ltimes T(2)$ in the case of the Poincaré group in Minkowski space, and $Q(2) \supset E(2) \cong SO(2) \ltimes T(2)$ in the Euclidean case[d]. For this it is necessary to be explicit as to the construction of the above-mentioned representation spaces. The 'scalar' case in the sense of Mackey's theory simply corresponds to setting $\mathbb{E}^{\mu}{}_{\nu} \equiv 0$, so that we are left with the well-known standard unitary irreducible representations of the two-dimensional Weyl-Heisenberg algebra for $\hbar > 0$. In the basis of Fock modes of a 'two-dimensional oscillator', the associated state space is precisely that occurring in the Schwinger[14] construction of representations of the $SU(2)$ angular momentum algebra, or alternatively used by Holman and Biedenharn[15] in the construction of unitary irreducible representations of $SU(1,1)$.

It remains to give the realisation of the above $Q(1,1)$ or $Q(2)$ generators in terms of Fock mode raising and lowering operators. An appropriate choice is

$$\begin{pmatrix} Z_1 \\ Z_2 \end{pmatrix} = \begin{pmatrix} a \\ b \end{pmatrix} = \begin{pmatrix} \frac{1}{\sqrt{2}}(X_1 + iP_1) \\ \frac{1}{\sqrt{2}}(X_2 + iP_2) \end{pmatrix}, \quad \begin{pmatrix} \overline{Z}_1 \\ \overline{Z}_2 \end{pmatrix} = \begin{pmatrix} a^{\dagger} \\ b^{\dagger} \end{pmatrix},$$

---

[c]Again, the statement extends easily to the $(D-1,1)$– and D–dimensional cases where the $\mathbb{E}^{\mu}{}_{\nu}$ generate $U(D-1,1)$ or $U(D)$, respectively.

[d]For work in related contexts see Refs. 16, 17.

for the Euclidean case, and

$$\begin{pmatrix} Z_0 \\ Z_1 \end{pmatrix} = \begin{pmatrix} a \\ b^\dagger \end{pmatrix} = \begin{pmatrix} \frac{1}{\sqrt{2}}(X_0+iP_0) \\ \frac{1}{\sqrt{2}}(X_1+iP_1) \end{pmatrix}, \quad \begin{pmatrix} \overline{Z}_0 \\ \overline{Z}_1 \end{pmatrix} = \begin{pmatrix} a^\dagger \\ b \end{pmatrix},$$

for the Minkowski case, respectively. These definitions then generate expressions for the remaining generators according to the definitions given above. For example, in the Euclidean case we have the $SO(2)$ generator $M := -iZ_{12} = -i(a^\dagger b - b^\dagger a)$, whereas in the Minkowski case the $SO(1,1)$ generator is $\Lambda := -iL_{01} = -i(a^\dagger b^\dagger - ab)$.

States are labelled as usual by standard mode numbers $|n_a, n_b\rangle$ relative to the zero-mode state $a|0,0\rangle = 0 = b|0,0\rangle$, namely

$$\begin{aligned} |n_a, n_b\rangle =& \frac{a^{\dagger n_a}}{\sqrt{n_a!}}\frac{b^{\dagger n_b}}{\sqrt{n_b!}}|0,0\rangle, \quad n_a, n_b = 0,1,2,\cdots; \\ N_a|n_a,n_b\rangle =& n_a|n_a,n_b\rangle; \qquad N_b|n_a,n_b\rangle = n_b|n_a,n_b\rangle; \quad \text{where} \\ N_a =& a^\dagger a, \qquad N_b = b^\dagger b. \end{aligned} \tag{4}$$

Convenient bases for the remaining generators of $U(2)$ and $U(1,1)$ in the respective cases are then as follows, and can easily be derived as linear combinations of the standard generators (2), (3) above:

$$U(2): \quad \begin{cases} J_+ = a^\dagger b \\ J_- = ab^\dagger \\ J_z = \frac{1}{2}(N_a - N_b) \\ U = N_a + N_b + 1 \\ M = -i(a^\dagger b - b^\dagger a) \end{cases} ; \qquad U(1,1): \quad \begin{cases} K_+ = a^\dagger b^\dagger \\ K_- = ab \\ K_0 = \frac{1}{2}N_a + \frac{1}{2}N_b + \frac{1}{2} \\ U = N_a - N_b \\ \Lambda = -i(a^\dagger b^\dagger - ab) \end{cases} \tag{5}$$

Finally, the Casimir operators of the homogeneous unitary groups are

$$\begin{aligned} C(SU(2)) =& J_z(J_z+1) + J_-J_+ \equiv \tfrac{1}{2}(N_a+N_b)\big(\tfrac{1}{2}(N_a+N_b)+1\big); \\ C(SU(1,1)) =& K_0(K_0-1) - K_+K_- \equiv \big(-\tfrac{1}{2}|N_a-N_b| - \tfrac{1}{2}\big)\big(-\tfrac{1}{2}|N_a-N_b| + \tfrac{1}{2}\big). \end{aligned} \tag{6}$$

The Fock mode basis (4) above is ideally adapted to the diagonalisation of the standard $U(1)$ compact generators J_z and K_0, respectively. Given (6), the eigenstates $|n_a, n_b\rangle$ simply give state labels within the following group branching chains:

$$\begin{aligned} Q(2) \supset U(2) \supset U(1): \quad & |j; m_z\rangle, \quad j = \tfrac{1}{2}(n_a+n_b),\ m_z = \tfrac{1}{2}(n_a - n_b); \\ Q(1,1) \supset U(1,1) \supset U(1): \quad & |j; n\rangle, \quad j = -\tfrac{1}{2}|n_a - n_b| - \tfrac{1}{4},\ n = \tfrac{1}{2}(n_a+n_b)+1. \end{aligned} \tag{7}$$

The situation is most easily seen graphically. In the figure is shown the spectrum of N_a, N_b plotted as a lattice of nodes along vertical and horizontal axes, respectively. Inspection of the examples given of states of fixed constant $n_a + n_b$ for the $U(2)$

case, and $|n_a - n_b|$ for the $U(1,1)$ case, is enough to show convincingly that our two dimensional Fock space carries the direct sum of representations of $SU(2)$ of spins $j = 0, \frac{1}{2}, 1, \frac{3}{2}, \cdots$ in the compact case as pointed out by Schwinger,[14] or of representations of $SU(1,1)$ in the positive discrete series D_j^+, with spins $j = -\frac{1}{2}, -\frac{3}{2}, -\frac{5}{2}, \cdots$ being doubly degenerate except for the case $j = -\frac{1}{2}$ (corresponding to the diagonal axis $n_a = n_b$), as pointed out by Holman and Biedenharn[15] (for notation see for example[18]).

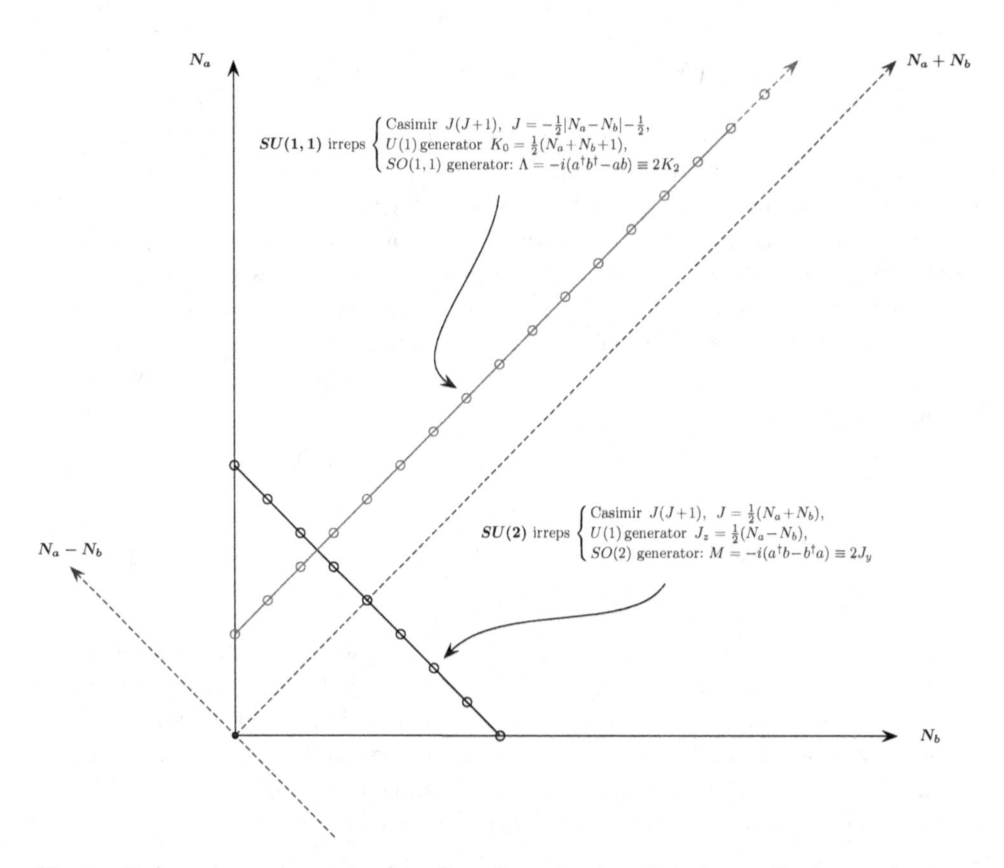

Fig. 1. Fock mode number states for a 'two dimensional oscillator' as a direct sum of unitary irreducible representations of $SU(2)$[14] or of $SU(1,1)$.[15]

Given the importance of the 'physical' Euclidean and Poincaré subalgebras of the quaplectic groups, we wish to consider the group reduction and state labelling problems associated with slightly different branching chains from (7). Namely, we wish to work *not* in the number basis directly, but instead with states diagonalised with respect to the relevant physical Lorentz group $SO(1,1)$ generated by Λ, or the physical symmetry group $SO(2)$ of rotations in the plane generated by M, respectively, where from (5) $\Lambda = iL_{01} = K_+ + K_- \equiv K_1$, $M = iL_{12} = J_+ + J_- \equiv J_y$. In the Euclidean case what we have chosen to call the '$U(1)$ versus the $SO(2)$' generators obviously differ only by a simple basis rotation which can be implemented[20] by introducing standard 'circular oscillators', and corresponding number states

$$
\begin{aligned}
a_\pm &= \tfrac{1}{\sqrt{2}}(a \mp ib), \qquad {a_\pm}^\dagger = \tfrac{1}{\sqrt{2}}(a^\dagger \pm ib^\dagger), \\
|n_+, n_-\rangle &= \frac{{a_+}^{\dagger n_+}}{\sqrt{n_+!}} \frac{a_-^{\dagger\, n_-}}{\sqrt{n_-!}} |0,0\rangle, \quad n_+, n_- = 0,1,2,\cdots; \\
M|n_+, n_-\rangle &= (N_+ - N_-)|n_+, n_-\rangle = (n_+ - n_-)|n_+, n_-\rangle,
\end{aligned}
$$

where as usual $N_\pm = a_\pm^\dagger a_\pm$. The decomposition in the basis of $\pm$ mode numbers thus reads identically to the decomposition given in the figure, where the $N_a - N_b$ axis is replaced by the $N_+ - N_-$ axis; note in particular that the eigenvalue j of the spin J remains $\frac{1}{2}(n_+ + n_-)$. However, the corresponding (unitary) basis transformation within $SU(1,1)$ is much more subtle, as it involves going from the compact generator K_0 to one of the two non-compact generators, K_1 and K_2. It is known[21,22] that in the positive discrete series D_j^+ the result is a direct integral of states $|j, \lambda\rangle$ with $-\infty < \lambda < \infty$. Thus starting from the basis (7) we now have the following symmetry-adapted bases,

$$
\begin{aligned}
Q(2) \supset SU(2) \supset SO(2): \quad & |j; m\rangle, \quad j = \tfrac{1}{2}(n_+ + n_-), \; m = (n_+ - n_-), \\
& \langle j; m | j'; m'\rangle = \delta_{jj'}\delta_{mm'}; \\
Q(1,1) \supset SU(1,1) \supset SO(1,1): \quad & |j; \lambda\rangle, \quad j = -\tfrac{1}{2}|n_a - n_b| - \tfrac{1}{2}, \; -\infty < \lambda < \infty, \\
& \langle j, \lambda | j', \lambda'\rangle = \delta_{jj'}\delta(\lambda - \lambda'). \qquad (8)
\end{aligned}
$$

The most natural symmetry-adapted bases for quaplectic representations are of course still not those of (8). Rather they should involve the physical spacetime symmetry groups embedded in $Q(2)$ and $Q(1,1)$. One possibility is of course simply the Fourier or 'momentum' bases, with diagonalised physical Euclidean translation and relativistic momentum generators, with the branching rule dictated by the spectrum of the relevant Casimir operators $P{\cdot}P$. Namely for $E(2)$ and $P(1,1)$ these bases are

$$
\begin{aligned}
Q(2) \supset T(2): \quad & P^a|\pi{\cdot}\pi; \pi^a\rangle = \pi^a|\pi{\cdot}\pi; \pi^a\rangle, \quad -\infty < \pi^1, \pi^2 < \infty; \\
Q(1,1) \supset T(2): \quad & P^\mu|p{\cdot}p; p^\mu\rangle = p^\mu|p{\cdot}p; p^\mu\rangle, \quad -\infty < p^0, p^1 < \infty; \qquad (9)
\end{aligned}
$$

with the spectrum of Euclidean and Poincaré group Casimirs, namely the appropriate invariant square of the Euclidean translation and relativistic momentum generators $P{\cdot}P = (P^1)^2 + (P^2)^2$, $P{\cdot}P = (P^0)^2 - (P^1)^2$, naturally being $0 < \pi{\cdot}\pi < \infty$, and $-\infty < p{\cdot}p < \infty$, respectively, simply reflecting the spectra of the individual components.

A final basis transformation – of importance for example in considering the $b \to \infty$ contraction limit of the quaplectic representations (see the introduction, and also the concluding remarks below) – entails the replacement of the unitary $SU(2)$ and $SU(1,1)$ groups in the previous labelling chains (8), by their space-time counterparts, the 'physical' Euclidean and Poincaré groups, respectively. Thus we seek to diagonalise, not the unitary group Casimir invariants, but the respective Euclidean and Poincaré group Casimirs, namely the appropriate invariant square of the Euclidean translation and relativistic momentum generators; while keeping diagonal the spatial rotations and spacetime Lorentz transformations (which of course commute with the momentum-squared Casimirs):

$$\begin{aligned}
Q(2) \supset E(2) \supset SO(2): \quad & |\pi\cdot\pi; m\rangle, \qquad m = 0, \pm\tfrac{1}{2}, \pm 1, \pm\tfrac{3}{2}, \cdots \\
& \langle \pi\cdot\pi; m|\pi'\cdot\pi'; m'\rangle = \delta(\pi^2 - \pi'^2)\delta_{mm'}; \\
Q(1,1) \supset P(1,1) \supset SO(1,1): \quad & |p\cdot p; \lambda\rangle, \quad -\infty < \lambda < \infty, \\
& \langle p\cdot p, \lambda|p'\cdot p', \lambda'\rangle = \delta(p^2 - p'^2)\delta(\lambda - \lambda').
\end{aligned} \tag{10}$$

Again, this time in both the Euclidean and Minkowski cases, this basis transformation corresponds to going from an operator with a discrete spectrum (the relevant unitary group Casimir), providing a countable basis, to a direct integral of eigenspaces of the non-compact momentum-squared operator. The construction is in fact simplest in the Fourier, or momentum basis, but can also be carried out via the unitary bases (8) in both cases. For example, in the Euclidean case, the $|\pi\cdot\pi; m\rangle$ states, in the position representation with two-dimensional polar coordinates (r, θ), are the well-known cylindrical eigenstates[20] $\propto J_{|m|}(r\sqrt{\pi\cdot\pi})e^{im\theta}$. $J_{|m|}$ is a standard circular Bessel function, and the action of the remaining unitary and quaplectic generators $Q(2) \sim E(2)$ such as $J_\pm$, $U = \frac{1}{2}(P_+P_- + X_+X_-)$, as well as $P_\pm = P^1 \pm iP^2$, $X_\pm = X^1 \pm iX^2$, on these can be given explicity (and the Heisenberg algebra recovered, as it must be). The Minkowski case can be handled similarly.

4. Conclusions

The idea of reciprocity has found resonance with various attempts to generalise the theoretical framework for the fundamental particle interactions – for example, in the guise of bi-crossproduct algebras and physics at the Planck scale,[23] "two-time" formulations,[24] or *ad hoc* "noncommutative geometry" extensions of perturbative field theory.[25] However, our discussion of branching rules and state labelling for unitary irreducible representations of the quaplectic group well illustrates the essential difficulty of any direct attempt to construct generalisations of spacetime Poincar'e symmetry. Namely, consistent with O'Raifeartaigh's theorem,[26,27] a *continuous*, *real* eigenvalue spectrum of the mass-squared Casimir operator $P^\mu P_\mu$ should obtain - a situation plainly not in accord with observation!

At the formal level however, it should be noted that quaplectic symmetry is often already implicit in any treatment of phase space for relativistic systems. For example, off-shell momentum integrations in field theory calculations in fact can be regarded as implementing a completeness relation for 'scalar' unitary irreducible representations of the quaplectic group $Q(3,1)$ or $Q(D-1,1)$. A further aspect of this purely formal role of quaplectic symmetry is that, since it contains the Poincaré algebra as a contraction limit (with the parameter $b \rightarrow \infty$), there should exist limits of quaplectic unitary irreducible representations yielding, for example, states of a massive scalar particle (solutions of the Klein-Gordon equation), much as non-relativistic Schrödinger wavefunctions (unitary irreducible representations of the Galilean group) can in turn be recovered in the $c \rightarrow \infty$ limit from solutions of the Klein-Gordon equation.[13]

For all the above reasons, and despite the difficulties posed by O'Raifeartaigh's theorem, there is a strong argument for taking Born reciprocity seriously. The study of the quaplectic symmetry group of reciprocal relativity is as important as, but has hitherto been neglected relative to, traditional extensions such as conformal and de Sitter symmetries. Recent approaches, such as the investigation of the implications of the uncertainty relations (Schrödinger-Robertson inequalities) in the context of the meaning of a 'semiclassical limit' of quantum quaplectic symmetry,[28] or indeed the possibility of formulating Hamiltonian worldline quantisations of reciprocally-invariant systems,[29] are initial steps in this direction.

Acknowledgements

The present work would not have been possible without ongoing collaboration and correspondence with Stuart Morgan[30] (Hobart), Jan Govaerts (Louvain) and especially Stephen Low (Austin), who has generously supplied details of working notes and draft papers on reciprocal relativity.

It is a pleasure to acknowledge more than twenty-five years of positive physics interactions and support across Bass Strait, between the theory group in Hobart at the University of Tasmania, and the Physics Department at the University of Melbourne. A constant beacon across the water and inspiration down the years has been the enthusiasm, activity and influence of Bruce McKellar and Girish Joshi, coupled with their high standards and achievements in many domains of theoretical physics, as attested to by the contributions at this meeting.

References

1. M. Born, *Elementary Particles and the Principle of Reciprocity*, *Nature* **163**, 207–208 (1949)
2. M. Born, *Reciprocity Theory of Elementary Particles*, *Reviews of Modern Physics* **21**(3), 463–473 (1949)
3. H. S. Green, *Quantized Field Theories and the Principle of Reciprocity*, *Nature* **163**, 208–209 (1949)
4. H. S. Green, *Theory of Reciprocity, Broken SU(3) Symmetry, and Strong Interactions*, Proc. Int. Conf. on Elementary Particles, Kyoto, 1965, *Prog. Theory. Phys.* (1966) 159
5. J. Bernstein, *Max Born and the quantum theory* , American J Phys **73**, 999–1008 (2005); Erratum, American J Phys **74**, 160 (2006)

6. E. Wigner, *On Unitary Representations of the Inhomogeneous Lorentz Group* , Ann Math **40**, 149–204 (1939)
7. Peter Szekeres, *Mathematical physics at the University of Adelaide* , Reports on Math Phys **57**, 3–11 (2006)
8. A. J. Bracken, *Group-theoretical applications in a tri-local model for baryons*, PhD thesis, University of Adelaide (1971)
9. Stephen G. Low, *Representations of the Canonical Group, (the Semidirect Product of the Unitary and Weyl-Heisenberg Groups), Acting as a Dynamical Group on Non-commutative Extended Phase Space*, *J. Phys. A* **35**, 5711–5729 (2002)
10. Stephen G. Low, *Poincaré and Heisenberg Quantum Dynamical Symmetry: Casimir Invariant Field Equations of the Quaplectic Group*, `math-ph/0502018`
11. Stephen G. Low, *Reciprocal Relativity of Noninertial Frames and the Quaplectic Group*, *Found. Phys.* **36** (7), 1036–1069 (2006), `math-ph/0506031`
12. P. A. M. Dirac, *Unitary Representations of the Lorentz Group*, *Proc. Roy. Soc. (London)* **A183**, 284 (1945)
13. E. J. Saletan, *Contraction of Lie Groups*, *J. Math. Phys.* **2**, 1–21 (1961); (E) *ibid* **2**, 742 (1961)
14. J. Schwinger, *On angular momentum*, in "Quantum Theory of Angular Momentum", Eds L. C. Biedenharn and H. van Dam, New York: Academic Press (1965)
15. W. J. Holman and L.C̃. Biedenharn, *Complex angular momenta and then groups* $SU(1,1)$ *and* $SU(2)$, *Ann Phys* **39**, 1–42 (1966)
16. J. W. B. Hughes, *Irreducible Representations of the central extension of* $Sl(2) \wedge T(2)$, *J Math Phys* **22** (12), 2775–2779 (1981)
17. R. J. B. Fawcett and A. J. Bracken, *Simple orthogonal and unitary non-compact quantum systems and the Inonu-Wigner contraction* , *J Math Phys* **29** (7), 1521–1528 (1988)
18. B. G. Wybourne, "Classical Groups for Physicists", New York:Wiley (1974)
19. J. A. Wolf, *Representations of certain semidirect product groups* , *J Funct Anal* **19**, 339–372 (1975)
20. B. H. Bransden and C. J. Joachain, "Introduction to Quantum Mechanics", New York:Longman (1989)
21. N. Mukunda, *Unitary representations of the group* $O(2,1)$ *in an* $O(1,1)$ *basis* , *J Math Phys* **8** (11), 2210–2220 (1967)
22. G. Lindblad and B. Nagel, *Continuous bases for unitary irreducible representations of* $SU(1,1)$, Annales de l'Institut Henri Poincaré **AXIII(1)**, 27–56 (1970)
23. S. Majid, *Hopf Algebras for Physics at the Planck Scale*, *Class. Quant. Grav.* **5**, 1587–1606 (1993)
24. See for instance, and references therein,
    I. Bars, *The Standard Model of Particles and Forces in the Framework of 2T-physics*, *Phys. Rev. D* **74**, 085019 (2006)
25. For references and recent discussions, see for example,
    A. Connes, "Noncommutative Geometry" London:Academic Press (1994);
    R. Oeckl, *Braided Quantum Field Theory*, *Comm. Math. Phys.* **217**, 451–473 (2001);
    V. Rivasseau, *Noncommutative Renormalization*, e-print `arXiv:0705.0705 [hep-th]`
26. L. O'Raifeartaigh, *Internal Symmetry and Lorentz Invariance*, *Phys. Rev. Lett.* **14**, 332 (1965);
    L. O'Raifeartaigh, *Mass Differences and Lie Algebras of Finite Order*, *Phys. Rev. Lett.* **14**, 575 (1965);
    L. O'Raifeartaigh, *Lorentz Invariance and Internal Symmetry*, *Phys. Rev.* **139**, B1052 (1965)

27. G. Fuchs, *About O'Raifeartaigh's Theorem*, *Ann. Inst. Henri Poincaré* **IX(1)**, 7–16 (1968)
28. P. D. Jarvis and S. O. Morgan, *Born Reciprocity and the Granularity of Spacetime*, *Found. Phys. Lett.* **19** (6), 501-517 (2006), e-print `arXiv:math-ph/0508041`
29. J. Govaerts, P. D. Jarvis, S. O. Morgan and S. G. Low, *Worldline quantisation of a reciprocally invariant system*, e-print `arXiv:0706.3736 [hep-th]`
30. S. O. Morgan, Ph D thesis, University of Tasmania (2007)

QUARK-LEPTON SYMMETRY AND QUARTIFICATION IN FIVE DIMENSIONS*

KRISTIAN L. MCDONALD[†]

Research Center for High Energy Physics, University of Melbourne, Victoria, 3010, Australia
[†]*E-mail: k.mcdonald@physics.unimelb.edu.au*

We outline some features of higher dimensional models possessing a Quark-Lepton (QL) symmetry. The QL symmetric model in five dimensions is discussed, with particular emphasis on the use of split fermions. An interesting fermionic geography which utilises the QL symmetry to suppress the proton decay rate and to motivate the flavor differences in the quark and leptonic sectors is considered. We discuss the quartification model in five dimensions and contrast the features of this model with traditional four dimensional constructs.

Keywords: Extra dimensions, split fermions, quark-lepton symmetry, quartification.

1. Introduction

The Standard Model (SM) of particle physics displays a clear asymmetry between quarks and leptons. Quarks and leptons have different masses and charges and importantly quarks experience the strong force. Despite these differences a suggestive similarity exists between the family structure of quarks and leptons, leading one to wonder if the SM may be an approximation to a more symmetric fundamental theory.

The similar family structure of quarks and leptons is an automatic consequence of the defining symmetry of the Quark-Lepton (QL) symmetric model.[1] In this framework the similarity between quarks and leptons is elevated to an exact symmetry of nature. The model permits a complete interchange symmetry between quarks and a generalized set of leptons, with the SM resulting from the breaking of an enlarged symmetry group.

On an independent front, a number of model building tools which employ extra spatial dimensions have been developed in recent years. In particular new methods of symmetry breaking have been uncovered. These methods allow one to reduce the scalar content required to achieve symmetry breaking in four dimensionsal models. It

*This talk is based on work completed in collaboration with A. Coulthurst, A. Demaria and B. H. J. McKellar.

is natural to investigate these methods within pre-existing frameworks to determine the phenomenological distinctions between the new and traditional approaches.

In this work we outline some recent investigations in five dimensional models possessing a QL symmetry. We focus our attention on two aspects of these investigations. First we consider the use of split fermions in models with a QL symmetry. We show that the symmetry constrains the implementation of split fermions and allows one to solve some outstanding problems in four dimensional with QL models; namely how can one resolve the notion of a QL symmetry with the disimilarity between the observed masses of quark and lepton.

We also consider the quartification model, a framework which extends the notion of a QL symmetry to permit gauge unification, in five dimensions. We show that the higher dimensional quartification model allows one to remove many of the complications which arise in four dimensional models. These complications result mainly from the relatively large Higgs sector required to achieve the necessary symmetry breaking. We show that an effective Higgsless limit may be obtained in the five dimensional quartification model, thereby permitting considerable simplification.

2. Quark-Lepton Symmetry

How does one construct a quark-lepton symmetric model? Let us recall that the SM also displays a clear asymmetry between left and right handed fields; the left handed fields experience $SU(2)_L$ interactions whilst the right handed fields do not. One generation of SM fermions may be denoted as

$$Q, L, u_R, d_R, e_R, \tag{1}$$

revealing a further left-right asymmetry; namely the absence of ν_R. However the left-right asymmetry may be a purely low energy phenomenon and the construction of a high energy left-right symmetric theory proceeds as follows. One must first extend the SM particle content to include ν_R and thereby equate the number of left and right degrees of freedom. The SM gauge group must also be enlarged:[2–5]

$$SU(3) \otimes SU(2) \otimes U(1) \to SU(3) \otimes [SU(2)]^2 \otimes U(1),$$

where the second $SU(2)$ group acts on right-chiral fermion doublets. This enables one to define a discrete symmetry interchanging all left and right handed fermions in the Lagrangian, $f_L \leftrightarrow f_R$. Furthermore the model must be constructed such that the additional symmetries introduced to enable the left-right interchange are suitably broken to reproduce the SM at low energies. In practise this means that a suitable extension of the Higgs sector is also required.

One may construct a quark-lepton symmetric model by employing the same recipe.[1,6–11] First the fermion content of the SM must be extended. As quarks come in three colors one is required to introduce more leptons to equate the number of quark and lepton degrees of freedom. For each SM lepton one includes two exotic leptons,

$$e \to E = (e, e', e''),$$
$$\nu \to N = (\nu, \nu', \nu''), \tag{2}$$

where the primed states are the exotics (known as liptons in the literature). The gauge group must also be extended:

$$SU(3) \otimes SU(2) \otimes U(1) \to [SU(3)]^2 \otimes SU(2) \otimes U(1),$$

where the additional $SU(3)$ gauge bosons induce transitions amongst the generalized leptons in the same way that gluon exchange enables quarks to change color. Now one may define a discrete symmetry interchanging all quarks and (generalized) leptons in the Lagrangian,

$$Q \leftrightarrow L, \quad u_R \leftrightarrow E_R, \quad d_R \leftrightarrow N_R. \tag{3}$$

The gauge symmetry may be broken via the Higgs mechanism to reproduce the charge differences between quarks and leptons and give heavy masses to the liptons. Phenomenologically consistent models can be obtained by breaking the lepton color group $SU(3)_\ell$ completely or by leaving an unbroken subgroup $SU(2)_\ell \subset SU(3)_\ell$ (which serves to confine the liptons into exotic bound states). The QL symmetry implies mass relations of the type $m_u = m_e$ which are more difficult to rectify. In 4D one may remove these by extending the scalar sector and thereby increasing the number of independent Yukawa couplings.

3. QL Symmetry in Five Dimensions

In recent years the study of models with extra dimensions has revealed a number of new model building tools. The use of orbifolds provides a new means of reducing the gauge symmetry operative at low energies by introducing a new mass scale, namely the compactification scale $M_c = 1/R$. In generic 5D models the mass of exotic gauge bosons can be set by M_c and the reduction of symmetries by orbifold construction results in collider phenomenology which differs from that obtained in models employing the usual 4D Higgs symmetry breaking.

The reduction of the QL symmetric gauge group via orbifold construction has recently been studied in 5D.[12] One question that faces the model builder in extra dimensional models is whether to place fermions on a brane or in the bulk. We shall focus on the latter in what follows and demonstrate that the troublesome mass relations which arise in 4D QL models actually provide useful and interesting model building constraints in 5D models.[13] We note that placing fermions in the bulk permits the unification of quark and lepton masses at TeV scales in more a general class of models.[14,15]

In five dimensions bulk fermions lack chirality. However, chiral zero mode fermions, which one may identify with SM fermions, may be obtained by employing orbifold boundary conditions on a fifth dimension forming an S^1/Z_2 orbifold. By coupling a bulk fermion to a bulk scalar field one may readily localise a chiral zero mode fermion at one of the orbifold fixed points.[16] Denoting the bulk fermion (scalar) as ψ (ϕ) one has the following Lagrangian:

$$\mathcal{L} = \bar{\psi}(\Gamma_M \partial^M - f\phi)\psi - \frac{1}{2}\partial_M \phi \partial^M \phi - V(\phi), \tag{4}$$

where Γ_M are the Dirac matrices, f is a constant, $M = 0, 1...4$ is the 5D Lorentz index and

$$V(\phi) = \frac{\lambda}{4}(\phi - v)^2, \tag{5}$$

is the usual quartic potential. If ϕ transforms trivially under Z_2 its ground state is given by $\langle\phi\rangle = v$. However if ϕ is odd under Z_2 its ground state is required to vanish at the fixed points. This results in a kink vacuum profile for ϕ which serves to localise chiral zero mode fermions, ψ_0, at one of the orbifold fixed points. The fixed point at which ψ_0 is localised depends on the sign of f.

In a QL symmetric model one must specify the transformation properties of the bulk scalar under the QL symmetry. An interesting choice is to make ϕ odd under the QL symmetry,[17] resulting in a Yukawa Lagrangian of the form

$$\mathcal{L} = -\left\{f_Q(Q^2 - L^2) + f_u(U^2 - E^2) + f_d(D^2 - N^2)\right\}\phi,$$

where $Q^2 = \bar{Q}Q$, etc, and the SM fermions are identified with the chiral zero modes of the bulk fermions in an obvious fashion. Note that the choice $f_Q, f_u, f_d > 0$ automatically implies the localisation pattern shown in Figure 1. It is of interest that this geography may be implemented in a less arbitrary fashion in a QL symmetric framework as this pattern is precisely that advocated recently to suppress the proton

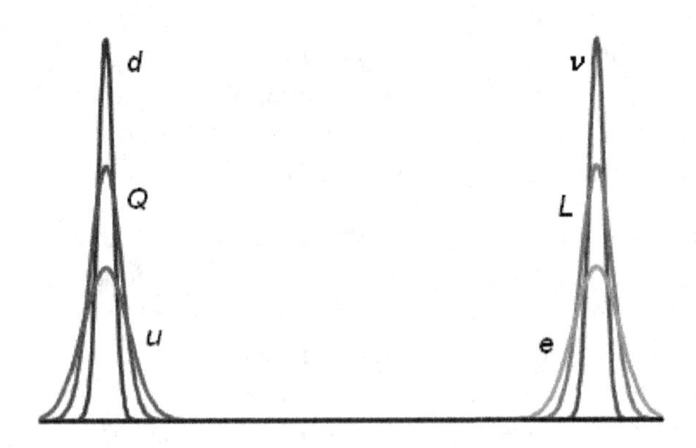

Fig. 1. The 5D wavefunctions for quarks and leptons in a QL symmetric model with one bulk scalar.

decay rate without an extremely large ultra-violet cutoff.[18] Indeed the effective 4D proton decay inducing non-renormalizable operator has the form

$$\mathcal{O}_p = \frac{K}{\Lambda^2} \mathcal{O}_Q^3 \mathcal{O}_L \tag{6}$$

where $\mathcal{O}_Q$ ($\mathcal{O}_L$) generically denotes a quark (lepton) operator and $K \sim \exp\left\{-vL^{3/2}\right\}$ represents the wavefunction overlap between quarks and leptons in the extra dimension. L denotes the length of the fundamental domain of the orbifold.

Observe that fermions related by the QL symmetry necessarily develop identical wavefunction profiles in the extra dimension and consequently the troublesome mass relations of the type $m_e = m_u$ persist in the effective 4D theory. With one bulk scalar it is only possible to localise fermions at the orbifold fixed points. However two bulk scalar models enable one to localise fermions within the bulk. This works as follows.[19,20] With one bulk scalar chiral zero mode fermions are found at one of the orbifold fixed points, with the precise point of localisation determined by the sign of the relevant Yukawa coupling. If a second bulk scalar is added with an opposite sign Yukawa coupling it will tend to drag the fermion towards the other end of the extra dimension, thereby localising it in the bulk. Importantly though the chiral fermion cannot be pulled very far into the bulk before it is dragged all the way to the other end of the extra dimension.

Let us add a second bulk scalar which is even under the QL symmetry, giving rise to the Lagrangian

$$\begin{aligned}\mathcal{L}_2 = -&\left\{h_Q(Q^2 + L^2) + h_u(U^2 + E^2)\right.\\ &\left.+h_d(D^2 + N^2)\right\}\phi'.\end{aligned}$$

Let us again require $h_Q, h_u, h_d > 0$ so that all quark Yukawa couplings are positive. Both ϕ and ϕ' will attempt to localize quarks at the same point in the extra dimension and they will remain at their original point of localisation. This type of quark geography is precisely that recently advocated[23] to allow one to construct quark flavor without introducing large flavor changing neutral currents. However ϕ and ϕ' attempt to localise leptons at different fixed points and the resultant point of localisation for a given lepton depends on which scalar dominates (these statements will be made numerically precise in a forthcoming publication[21]). An arrangement typical of this setup is shown in Figure 2.
This arrangement has a number of interesting features. Firstly note that shifting leptons into the bulk significantly alters the degree of wavefunction overlap in the quark and leptonic sectors. When one obtains the effective 4D fermion masses the different degree of overlap in the quark and lepton sectors will remove the undesirable QL mass relations. Furthermore the overlap between left and right chiral fermion wavefunctions is expected to be greater in the quark sector than in the lepton sector, leading to the generic expectation that quarks will be heavier than the lepton to which they're related by the QL symmetry. Observe that dragging the

right chiral neutrinos ν_R all the way to the 'quark end' of the extra dimension significantly suppresses the neutrino Dirac masses below the electroweak scale. Dragging ν_R to the quark end of the extra dimension does not introduce rapid proton decay, provided that ν_R has a large enough Majorana mass to kinematically preclude decays of the type $p \to \pi\nu_R$. Such a mass is expected to arise at the non-renormalizable level even if it is not induced by tree-level couplings in the theory.[22]

Thus the non-desirable Yukawa relations implied by 4D QL symmetric models turn out to be of interest in the 5D construct. They enable one to understand proton longevity within the split fermion framework in a less arbitrary fashion and instead of inducing unwanted mass relations suggest an underlying motivation for the flavour differences experimentally observed in the quark and lepton sectors.

4. Quartification in Five Dimensions

Adding a leptonic color group to the SM clearly renders the traditional approaches to gauge unification inapplicable. Recall that the simplest grand unified theory, namely $SU(5)$, does not contain the left-right symmetric model. However there are larger unifying groups which do admit the left-right symmetry, for example $SO(10)$ and the trinification gauge group $[SU(3)]^3 \times Z_3$. Similarly it is possible to construct a unified gauge theory admitting the QL symmetry by considering larger unification groups. It is natural to extend the notion of trinification to include a leptonic color factor, leading one to the so called quartification model. This possesses the gauge group $G_Q = [SU(3)]^4 \times Z_4$, where the additional $SU(3)$ factor corresponds to lepton color and the Z_4 symmetry cyclicly permutes the group factors to ensure a single coupling constant.[24–29]

It was been demonstrated that unification may be achieved within the quartification framework in 4D by enforcing additional symmetries upon the quartification model.[25] Subsequent work has shown that unification need not require additional symmetries, but does require multiple symmetry breaking scales between the unifi-

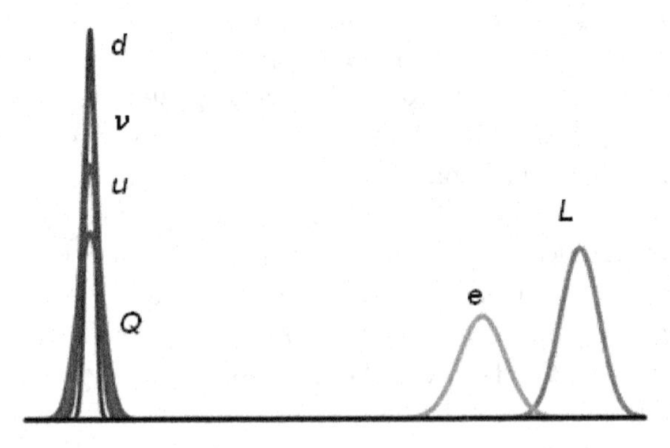

Fig. 2. The 5D wavefunctions for quarks and leptons in a QL symmetric model with two bulk scalars.

cation scale and the electroweak scale.[27] It was also shown[27] that unification may be achieved via multiple symmetry breaking routes both with and without the remnant leptonic color symmetry $SU(2)_\ell$.

The necessary symmetry breaking is accomplished with eight Higgs multiplets, giving rise to a complicated Higgs potential with a large number of free parameters. The demand of multiple symmetry breaking scales also requires hierarchies of vacuum expectation values (VEV's) to exist within individual scalar multiplets, giving rise to a generalized version of the doublet-triplet splitting problem familiar from $SU(5)$ unified theories. A large number of electroweak doublets also appear in the 4D constructs. Thus many of the less satisfactory features of 4D quartification models revolve around the Higgs sector required for symmetry breaking.

Recently the quartification model has been studied in 5D, where intrinsically higher dimensional symmetry breaking methods exist.[29] By taking the fifth dimension as an $S^1/Z_2 \times Z_2'$ orbifold one may employ orbifold boundary conditions (OBC's) on the bulk gauge sector to reduce the gauge symmetry operative at the zero mode level from G_Q to

$$SU(3)_c \otimes SU(2)_L \otimes SU(2)_\ell \otimes SU(2)_R \otimes U(1)^3. \tag{7}$$

However the use of OBC's does not reduce the rank of the gauge group so that further symmetry breaking is required. Rank reducing symmetry breaking can be achieved in higher dimensional theories by employing a boundary scalar sector to alter the boundary conditions on the compactified space for gauge fields.[30] Denoting a boundary scalar as χ and defining $V \propto \langle\chi\rangle$ one can show that V induces a shift in the Kaluza-Klein mass spectrum of gauge fields which couple to χ. If such a gauge field initially possessed a massless mode its tower receives a shift of the form

$$M_n \approx M_c\,(2n+1)\left(1+\frac{M_c}{\pi V}+\dots\right), \quad n = 0, 1, 2, \dots \tag{8}$$

giving a tower with the lowest-lying states M_c, $3M_c$, $5M_c$, ... This represents an offset of M_c relative to the $V = 0$ tower, with the field no longer retaining a massless zero mode. The association of V with the VEVs of the boundary scalar sector implies that the limit $V \to \infty$ is attained when $\langle\chi\rangle \to \infty$. However, when the VEVs of the Higgs fields are taken to infinity, the shift in the KK masses of the gauge fields is finite, giving the exotic gauge fields masses dependent only upon the compactification scale M_c. Consequently, these fields remain as ingredients in the effective theory while the boundary Higgs sector decouples entirely, and we can view our reduced symmetry theory in an effective Higgsless limit. Interestingly, in this limit also, the high-energy behaviour of the massive gauge boson scattering remains unspoilt as shown in Ref. 30.

It was shown in Ref. 29 that a unique set of OBC's was required to ensure that quark masses could be generated and to prevent liptons from appearing at the electroweak scale. The inclusion of a boundary Higgs sector allows one to reduce the quartification gauge symmetry down to the SM gauge group G_{SM} or to $G_{SM} \otimes SU(2)_\ell$ at the zero mode level. In both cases fifth dimensional components of the $SU(3)_L$ gauge fields with the quantum numbers of the SM Higgs doublet retained a massless mode, enabling one to use Wilson loops to reproduce the SM flavour structure.

A surprising result however was that unification could only be achieved when the remnant lepton color symmetry $SU(2)_\ell$ remained unbroken. Thus one arrives at a unique minimal quartification model which unifies in 5D, a result to be contrasted with the 4D case where a large number of symmetry breaking routes which permit unification have been uncovered. Unfortunately the unifying case requires the compactification scale to be greater than 10^{10} GeV so that only a SM like Higgs field is expected to appear at the LHC.

5. Conclusion

Quark-lepton symmetric models in some sense unify the fermionic content of the SM and thereby motivate the similar family structures observed in the quark and lepton sectors. Recent investigations involving QL symmetries in 5D have uncovered a number of interesting results. In particular the QL symmetry provides useful Yukawa relationships in split fermion models and the quartification model is found to be more constrained in 5D. A number of avenues for further investigation remain, with a detailed analysis of the fermionic geography of Figure 2 required.[21] It would also be interesting to combine the QL symmetry with a left-right symmetry in 5D, enabling one to simultaneously employ results uncovered in each of these frameworks.[31]

References

1. R. Foot and H. Lew, *Phys. Rev. D* **41** (1990) 3502.
2. J. C. Pati and A. Salam, *Phys. Rev. D* **10**, 275 (1974).
3. R. N. Mohapatra and J. C. Pati, *Phys. Rev. D* **11**, 566 (1975).
4. R. N. Mohapatra and J. C. Pati, *Phys. Rev. D* **11**, 2558 (1975).
5. G. Senjanovic and R. N. Mohapatra, *Phys. Rev. D* **12**, 1502 (1975).
6. R. Foot and H. Lew, *Phys. Rev. D* **42** (1990) 945.
7. R. Foot and H. Lew, *Mod. Phys. Lett. A* **5**, 1345 (1990).
8. R. Foot, H. Lew and R. R. Volkas, *Phys. Rev. D* **44**, 1531 (1991).
9. Y. Levin and R. R. Volkas, *Phys. Rev. D* **48**, 5342 (1993) [arXiv:hep-ph/9308256].
10. D. S. Shaw and R. R. Volkas, *Phys. Rev. D* **51**, 6490 (1995) [arXiv:hep-ph/9410350].
11. R. Foot and R. R. Volkas, *Phys. Lett. B* **358**, 318 (1995) [arXiv:hep-ph/9505331].
12. K. L. McDonald and B. H. J. McKellar, *Phys. Rev. D* **74**, 056005 (2006) [arXiv:hep-ph/0609110].
13. A. Coulthurst, A. Demaria, K. L. McDonald and B. H. J. McKellar, arXiv:hep-ph/0611269.
14. P. Q. Hung, Nucl. Phys. B **720** (2005) 89 [arXiv:hep-ph/0412262].
15. M. Adibzadeh and P. Q. Hung, arXiv:0705.1154 [hep-ph].
16. H. Georgi, A. K. Grant and G. Hailu, *Phys. Rev. D* **63**, 064027 (2001) [arXiv:hep-ph/0007350].

17. A. Coulthurst, K. L. McDonald and B. H. J. McKellar, *Phys. Rev. D* **74**, 127701 (2006) [arXiv:hep-ph/0610345].
18. N. Arkani-Hamed and M. Schmaltz, *Phys. Rev. D* **61**, 033005 (2000) [arXiv:hep-ph/9903417].
19. Y. Grossman and G. Perez, *Phys. Rev. D* **67**, 015011 (2003) [arXiv:hep-ph/0210053].
20. Y. Grossman and G. Perez, *Pramana* **62**, 733 (2004) [arXiv:hep-ph/0303243].
21. A. Coulthurst and K. L. McDonald, in preparation.
22. A. Coulthurst, K. L. McDonald and B. H. J. McKellar, *Phys. Rev. D* **75**, 045018 (2007) [arXiv:hep-ph/0611164].
23. B. Lillie, *JHEP* **0312**, 030 (2003) [arXiv:hep-ph/0308091].
24. G. C. Joshi and R. R. Volkas, *Phys. Rev. D* **45**, 1711 (1992).
25. K. S. Babu, E. Ma and S. Willenbrock, *Phys. Rev. D* **69**, 051301 (2004) [arXiv:hep-ph/0307380].
26. S. L. Chen and E. Ma, *Mod. Phys. Lett. A* **19**, 1267 (2004) [arXiv:hep-ph/0403105].
27. A. Demaria, C. I. Low and R. R. Volkas, *Phys. Rev. D* **72**, 075007 (2005) [arXiv:hep-ph/0508160].
28. A. Demaria, C. I. Low and R. R. Volkas, *Phys. Rev. D* **74**, 033005 (2006) [arXiv:hep-ph/0603152].
29. A. Demaria and K. L. McDonald, Phys. Rev. D **75**, 056006 (2007) [arXiv:hep-ph/0610346].
30. C. Csaki, C. Grojean, H. Murayama, L. Pilo and J. Terning, *Phys. Rev. D* **69**, 055006 (2004) [arXiv:hep-ph/0305237].
31. A. Coulthurst, J. Doukas and K. L. McDonald, arXiv:hep-ph/0702285.

MSW IN REVERSE; WHAT SNO SAYS ABOUT PSEUDO-DIRAC NEUTRINOS

G. J. STEPHENSON, JR.

Department of Physics and Astronomy, UNM,
Albuquerque, New Mexico 87131, USA
**E-mail: ab_gjs@phys.unm.edu*

I discuss the implications of the SNO neutral current result for the allowed Δm^2 values of pseudo-Dirac pairs. In particular, for certain parameter ranges, adiabatic MSW may occur in reverse as solar neutrinos traverse the earth's atmosphere.

Keywords: Neutrinos.

1. Introduction

It is a great pleasure to be here and to join in the honoring of two old friends. Bruce and Girish have made many contributions to particle physics in a number of subfields. Today, I want to speak about possibility for neutrino states that Bruce has been very instrumental in elucidating. This is meant to be a conversation and is very short on any derivations, they appear elsewhere in the literature. The citations are few and usually to old papers, which necessarily leaves out a great body of work over the past several decades as this is not meant to be a comprehensive review.

Conventional wisdom now describes the deficit of ^8B neutrinos from the sun as being due to an adiabatic Landau-Zener effect arising from the MSW process in the solar interior.[1] As will be discussed below, currently favored parameter values suggest that this affect the high energy portion of the ^8B spectrum and not the lower energy $p-p$ neutrinos. Those neutrinos so affected will emerge from the sun in a mass eigenstate and propagate to the earth without further oscillation. Following Wolfenstein's original suggestion,[2] the discussion turns on the charged current interaction on active neutrino states, since any neutral current effects will be diagonal in this basis.

There are, however, some models of neutrino mass and mixing that include what is known as pseudo-Dirac neutrinos.[3] These will be defined below, but may be described as a nearly degenerate pair of Majorana mass eigenstates which are each nearly 50 per cent active and 50 per cent sterile where the active components are some linear combination of possible active states. For such pairs, the neutral current interaction may have serious consequences.

As the SNO experiment only allows for the total sterile component of solar neutrinos to be a few per cent, there is a real concern that pseudo- Dirac pairs may already be ruled out. It is this issue that I shall discuss today.

2. Standard Solar MSW

For this discussion, assume the existence of three active light neutrinos which could be Majorana or pure Dirac neutrinos. The observation of oscillations among interaction eigenstates (hereafter called flavor states) implies first, that at least some neutrinos have mass, and, second, that the flavor states are linear combinations of the mass eigenstates. The discussion is carried out in the rest frame of the sun, so the time evolution of the mass eigenstates is governed by a phase factor $e^{\imath mt/\gamma}$ where γ is the Lorentz time dilatation factor, $\gamma = E/m$. Since all neutrino masses are tiny compared to the energies required for them to be observable, they move at essentially the speed of light. Using that to convert time to length, the phase factor becomes $e^{\imath 1.27m^2 L/E}$, with m^2 in $(eV)^2$ and L/E in meter per MeV or kilometers per GeV.

The MSW effect takes cognizance of the fact that, in dense matter (where dense is a relative term as will be seen below) the weak interaction will modify that phase factor in ways that have consequence for earthbound experiments measuring particular neutrino fluxes from the sun. The effect of the weak charged current interactions on forward scattering amplitudes only depends on exchange terms, so a given flavor state is only affected by the density of charged leptons of the same flavor. In the sun there are electrons, but no muons or tauons, so only the ν_e components of the mass eigenstates are affected. In the frame of the sun, the additional phase factor becomes $e^{\imath G_F \rho_e E}$ where G_F is the Fermi coupling constant and ρ_e is the number density of electrons.

The neutral current forward scattering depends instead on $G_F \rho_N E$, where ρ_N is the number density of neutrons (protons and electron contribute equal and opposite amplitudes). This effects all flavors equally, hence contributes only a common phase which can be ignored when considering, as here, only active neutrinos.

If either E or ρ_e is small enough, the additional phase will have no essential effect, being equivalent to a slight change in L. If, however, $G_F \rho_e E$ is large enough, the two closest eigenstates will be re-diagonalized into one which has all of the ν_e content and an orthogonal state. As the neutrino propagates out from the center of the sun, the conditions change adiabatically, so the propagating state follows the highest energy eigenstate and emerges from the sun as a mass eigenstate. This mass eigenstate propagates to earth where its interaction in a detector reflects whatever mixture of flavor states it had when it exited the sun.

Given the electron density at the center of the sun and the energy spectrum of ^8B neutrinos, this happens for those Boron 8 neutrinos which should provide the bulk of the signal in the chlorine detectors that first observed a solar neutrino deficit. On the other hand, pp neutrinos have too low an energy, hence propagate as the mixture of mass eigenstates with which they were created.

This leads to the conventional picture of mass and mixing that requires one closely spaced doublet containing most of the electron flavor with a third mass eigenstate consisting of muon and tauon flavors. The absolute mass scale is not determined. Solar neutrinos require that the doublet lead to Δm^2 about $7 \times 10^{-5}(eV)^2$ with the third sufficiently removed that, combined, with the average of the doublet, gives rise to the atmospheric requirement of Δm^2 about $10^{-3}(eV)^2$. These two phenomena cannot determine whether the doublet lies below the singlet (so-called normal hierarchy) or above the singlet (so-called inverted hierarchy).

3. Pseudo-Dirac Neutrinos

By a pseudo-Dirac neutrino, following Wolfenstein,[4] I mean a pair of Majorana neutrino mass eigenstates separated by a very small mass difference. In the limit that the mass difference goes to 0, the pair should combine to form the bispinor associated with a normal Dirac particle, capable of carrying a conserved charge (*i. e.* particle number.) As Wolfenstein[4] pointed out, this means that the two Majorana mass eigenstates must each be fifty-fifty mixtures of active and sterile spinors, phased in such a way that their individual contributions to lepton number violating processes, such as neutrinoless double beta decay[5] tend to cancel, restoring lepton number conservation in the limit of equal masses.

4. Implications of SNO Data for Pseudo-Dirac Neutrinos

Before addressing the real problem, let me consider a toy problem. Suppose that we have a pseudo-Dirac neutrino whose active components only contain muon and tauon flavors. Then there is no charged current effect and the neutral current effect on the active component will be independent of the flavor mixture. However, there is no interaction with the sterile component. In the limit that this pseudo-Dirac pair was degenerate (a Dirac neutrino), in the presence of matter the propagating eigenstates would be purely active and purely sterile.
If the pair is truly pseudo-Dirac in the sense that in vacuum there is some small mass difference (a possible example is pseudo-Dirac pairs resulting from a rank one see-saw as discussed with Garbutt, Goldman and McKellar[3]), then, when $G_F \rho_N E$ is larger than the mass difference squared, the same effect takes place. Then, if the active state is produced in matter, it will propagate as active through sufficiently dense matter but if it exits into vacuum adiabatically only one member of the pair will propagate.

For solar neutrinos, this effect would exist along with the usual charge current effect, if the members of the doublet were really pseudo-Dirac pairs. In the rank one model mentioned before, the natural pattern is one purely active neutrino with an extremely small mass and two closely spaced pseudo-Dirac pairs with the pseudo-Dirac splitting much much less than the spacing between the pairs. This is an example of an inverted hierarchy. In this case, with all of the restrictive assumptions made here, I would expect those neutrinos propagating from the sun to be only one mass eigenstate of each pseudo-Dirac pair, implying that the neutrinos arriving from the sun would have a very large sterile component. This is ruled out by SNO!

5. Ways Out

Do these simple arguments really rule out any possibility for pseudo-Dirac pairs? The answer must depend, at least, on the size of the mass splitting. It also depends on the validity of the assumptions made above. I shall describe two, out of probably many, ways around this apparent prohibition.

5.1. *Fluctuations and vacuum oscillations*

Alex Friedland[6] has discussed the effects of severe density fluctuations found in most models of supernovae on neutrino propagation. Since the relevant parameter is $(\delta\rho/\rho)$, one might worry that, at the very edge of a density distribution, even for a stable star like the sun, this could become important. This clearly depends on the size of Δm^2 which sets the density scale at which vacuum propagation effectively begins, and for pseudo-Dirac neutrinos with extremely small Δm^2 one expects that such fluctuations will essentially reset to an active neutrino composed of a coherent combination of both members of the pair.

If this occurs, the propagation to the earth will undergo the standard vacuum oscillation which has long been considered with respect to the charged current solar deficit. The analysis is the same and leads to the requirement that Δm^2 is less than about $10^{-11}(eV)^2$.

5.2. *Reverse MSW*

Suppose for a moment that Ray Davis had been able to set up his chlorine detector at the center of an object which had the same density profile as the sun. (It would, of course, have to be another sun, but this a gedanken exercise.) In that case, assuming adiabatic density changes, all of the evolution from the initial ν_e to the object propagating in free space would be exactly undone and Davis would have observed a full strength signal. (As an aside, there then would have been no "Solar Neutrino Puzzle" and much of the experimental progress in neutrino physics might not have occurred.) In fact, for the Δm^2 appropriate to the solar neutrino solution, ρ_e has to be associated with about 100 gm/cc of Hydrogen to be relevant.

For the much smaller Δm^2 associated with a pseudo-Dirac pair, even densities encountered on earth could be important. Bruce taught me the fascinating fact that the average density of the sun and the density of the earth's atmosphere at sea level are both 1. For the former, the units are gm/cc, for the latter kilograms per cubic meter. This implies that, for Δm^2 less than $10^{-10}(eV)^2$, the inverse MSW effect will restore the pseudo-Dirac neutrino to an active state, evading the SNO restriction.

6. Conclusion

In this talk, I have focussed on the possibility of pseudo-Dirac neutrinos and on possible limits that may be put on them. While there may be other scenarios with sterile neutrinos (or, more properly, nearly sterile neutrinos with very small mixing with active neutrinos, pseudo-Dirac neutrinos arise naturally in work that Bruce has done with several of us in various collaborations that has looked at the possibility of modified see-saw schemes in which the mass matrix in the sterile sector has rank less than three. In rank 1 schemes, it is very natural to produce two pseudo-Dirac pairs with a splitting approximately given by m_D^2/M where m_D is given by the scale of Dirac neutrino masses and M is the scale of the mass in the sterile sector. This leads to Δm^2 of the order of m_D^3/M, for m_d of the order of $10^{-1}eV$ and M of the order of $1GeV$, Δm^2 is of the order of $10^{-12}(eV)^2$. Many see-saw schemes argue for much larger values of M, leading to even smaller values for Δm^2.

I have argued here that such pseudo-Dirac neutrinos are not ruled out by current experiments.

References

1. A. Yu. Smirnov, *Phys. Scripta* **T121**, 57 (2005); arXiv:hep-ph/0412391; see SNO collaboration, *Phys. Rev. Lett.* **89**, 011302 (2002); arXiv:nucl-ex/0204009, and earlier references contained therein.
2. L. Wolfenstein *Phys. Lett. B* **107**, 77 (1981).
3. G. J. Stephenson, Jr.,T. Goldman, B. H. J. McKellar and M. Garbutt, *Int. J. Mod. Phys. A* **20**, 6373 (2005).
4. L. Wolfenstein, *Nucl. Phys. B* **186**, 147 (1981).
5. W. C. Haxton and G. J. Stephenson, Jr., *Prog. Part. Nucl. Phys.* **12**, 409 (1984) (ed. Sir Denys Wilkinson).
6. A. Friedland and A. Gruzinov, arXiv:astro-ph/0607244.

LOOKING FORWARD TO THE LARGE HADRON COLLIDER

GEOFFREY N. TAYLOR[†]
School of Physics,
The University of Melbourne,
Victoria 3010, Australian
[†]*E-mail: g.taylor@physics.unimelb.edu.au*

In this paper an overview of the Large Hadron Collider program and status is given, including a brief description of the scientific background from which this ambitious program evolved. The emphasis is on the status of the Standard Model Higgs Boson, searches for which are the key component of the LHC program. A description of the ATLAS one of the two large general purpose experiments designed to detect evidence for the Higgs Boson and other data of interest to searches for physics beyond the standard model.

Keywords: LHC; ATLAS.

1. Introduction

Over the past two decades, with increasing accuracy and sophistication, measurement after measurement has been unable to find any significant discrepancies with the standard model (SM). The range of energies from eV atomic processes to the highest energy processes available at accelerators, using protons, electrons, muons and neutrinos, have all relentlessly returned results consistent with the SM predictions. And yet it is well documented that the SM cannot be the final description of nature. The model itself has many parameters that must be obtained from experiment, a limitations theorists would like to reduce by embedding the theory of the SM into a more general picture, where the number of parameters can be reduced via relationships inherent in models based upon higher symmetries.

Key amongst the parameters that are not provided by the SM are the masses of most of the fundamental particles of the theory (the zero mass photon and gluons being the only exceptions). The *standard* resolution of the mass problem in the SM is via the Higgs mechanism,[10] invoking the existence of an all pervading scalar field with proscribed properties to generate a symmetry breaking in the vacuum ground state, whilst maintaining the basic symmetries of the Lagrangian describing the electro-weak interaction. The interaction of the field with the particles of matter generates their masses breaking the underlying symmetry spontaneously, whilst the self-interaction of the field generates as as yet undiscovered particle, the Higgs

Boson. The SM has nothing to say about gravity, a clear short coming for any candidate *Theory of Everything*, the ultimate pursuit of further developments of the SM.

The separation between the energy scale of symmetry breaking of the weak interaction ($10^{2-3}GeV$), the potential Grand Unification scale ($10^{15-16}GeV$) and the Planck scale appropriate for quantum gravity ($10^{19}GeV$) (leading to the Hierarchy Problem) is not understood, albeit with many proposed explanations. In fact, closer to experimental reach is the need to resolve what new physics is required to restore unitarity in predictions for processes such as WW scattering. Without additional physics processes to damp the cross-section, infinities develop in the predictions for such processes, at the TeV energy scale.

In recent years that neutrinos have a non-zero mass has been confirmed with clear measurements of neutrino oscillations. Helicity considerations mean even a small neutrino mass results in right-handed neutrinos, which are currently not part of the SM. The question of whether neutrinos are Majorana (neutrinos are their own anti-particles) or Dirac in nature, as yet unresolved, will determine whether right-handed neutrinos are explicitly required in the SM. Also of note are results recently published by the MiniBooNE Collaboration[11] that remove the tantalising need for some additional *sterile* neutrinos.

Although the missing ingredients of a future, more complete SM provide essential directions for forthcoming experiments and give impetus to further theoretical development and phenomenology, perhaps the most surprising and far reaching results affecting the directions of particle physics come from astrophysics and cosmology [1]. The existence of dark matter and now dark energy, represent a challenge and an opportunity for particle physics.

2. Successes of the Standard Model

Since the discovery of the weak neutral current in neutrino scattering experiments at CERN with the Gargamelle bubble chamber,[12] the unified electro-weak theory has prevailed. With interaction Lagrangian based upon the SU(2) × U(1) symmetry[2–4] and a unifying relation between the couplings for each of the two sectors, the SM of electro-weak interactions has passed all tests thrown at it. In all processes involving photon-exchange, a parallel process with the exchange of the neutral weak gauge Z-boson also contributes with a relative strength specified via the coupling ratio (and the Z-mass for which the measured value must be used). The mass of the Z results in a strong energy dependence of the electro-weak predictions, but only as a kinematic effect. The basic mixing of the processes is set by the theory.

To relate the electro-weak theory to physical processes, the matter particles, quarks and leptons must be introduced, and their transformation properties under the SU(2) × U(1) must be specified. Assigning the simplest group representations to the three generations of quarks and leptons results in the SM. Quarks are assigned to left- and right-handed doublets, neutral and charged leptons are assigned to left-handed doublets and right-handed charged leptons are singlets under SU(2), as shown below. With the transformation of these multiplets under the SU(2) × U(1) symmetry provided via group theory, predictions for all fundamental electro-weak processes are unique and testable. The outstanding success of the SM model is that all such tests have confirmed it. No evidence indicating a need to go beyond the SM yet exists.

It should be noted that the SM is completed with the addition of the colour symmetry group SU(3) describing the strong interactions between quarks and gluons - QCD. These interactions are flavour blind. Each quark flavour is replicated in three colours, representing colour triplets, whilst the gluons are part of the SU(3) octet representation under colour. (Although QCD jet production will be an important characteristic of interactions at the LHC, it will not be described in this paper.)

Figure 1 shows the current status of key experimental measurements normalised to the appropriate SM expectation, in units of standard deviations, of each measurement from the expectation obtained from the fit to the SM of all the measurements. What is clear from this data is consistency of the data from a wide range of processes and energies to the SM.

3. The Standard Model Higgs

Detailed measurements of the W-mass at LEP[5] and the Tevatron collider[14] , together with recent improved measurements of the top-quark mass[15] provide further agreement with the SM, when the missing ingredient of the Higgs boson is included in the calculations. This impressive consistency is reason enough to expect to find the Higgs Boson or something else that can play a similar role.

Based upon quantum corrections the current measurements of top and W masses require a Higgs (or similar particle) with a mass less that $144 GeV/c^2$ (95% upper CL). The preferred value for the SM Higgs mass is $m_{\mathrm{H}} = 76^{+33}_{-24} GeV/c^2$ is lower than previous values due to the new combined analysis of the top-quark mass[15] with the value $m_t = 170.9 \pm 1.1(stat) \pm 1.5(syst) GeV/c^2$. As it turns out, sensitivity of SM model quantum corrections are more sensitive to the mass of the top-quark than to the mass of the Higgs. LEP data, in particular, was sensitive to the existence of the top quark before its discovery at Fermilab's Tevatron. LEP analysis yielded a best estimate of $\sim 175 GeV/c^2$ for the top mass, very close to the actual value. That experience emboldens precision measurement predictions for the Higgs mass. Direct Higgs searches carried out at in the final months of LEP rule out a SM Higgs below $114 GeV/c^2$, pushing the upper limit higher to $182 GeV/c^2$.[5]

Recent SM Higgs searches and a combined analyses by D0 and CDF[14] have set Higgs production cross-section upper limits to within an order of magnitude of that expected for a Higgs mass of $115 GeV/c^2$ and to within a factor 3.8 at $160 GeV/c^2$.

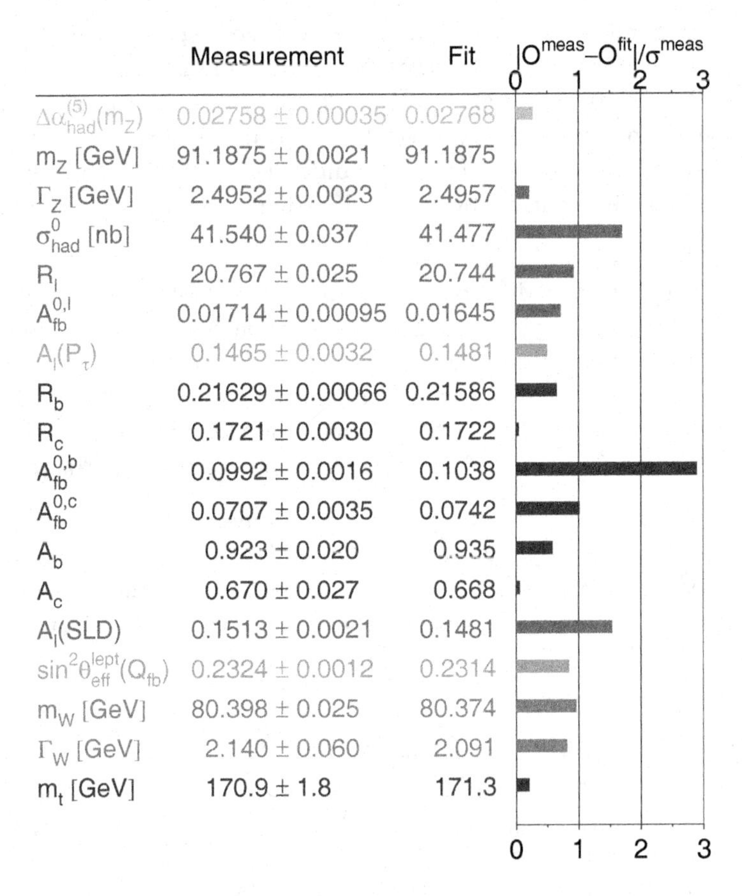

Fig. 1. Key electro-weak parameters from a variety of measurements and their deviations from values obtained by fitting all the data to the SM predictions. Each deviation is normalised by the measured error.[5]

These results are shown in Fig. 3. Tantalizingly close and with more data taking scheduled before the LHC enters production data-taking, Tevatron scientists are pushing hard to extract the most from current and future data.
Theoretical guidance also provides significant constraints on the expected Higgs mass. For example the calculation of the cross section for elastic scattering of longitudinally polarized W- or Z-bosons, including contributions from Higgs boson intermediate states, results in a non-physical breaking of the unitarity bound unless the Higgs mass is less than $800 - 1000 GeV/c^2$.[16]

Furthermore, considering the the Higgs self-interaction term that arises as part of the Higgs prescription. The Higgs mass runs with this coupling, resulting in an upper bound for the energy to which the theory is consistent. The self-interaction will become infinite is no bound is provided, unless it is identically zero - the *trivial* case, with a non-interacting field. The maintain a non-trivial interaction, there must be an

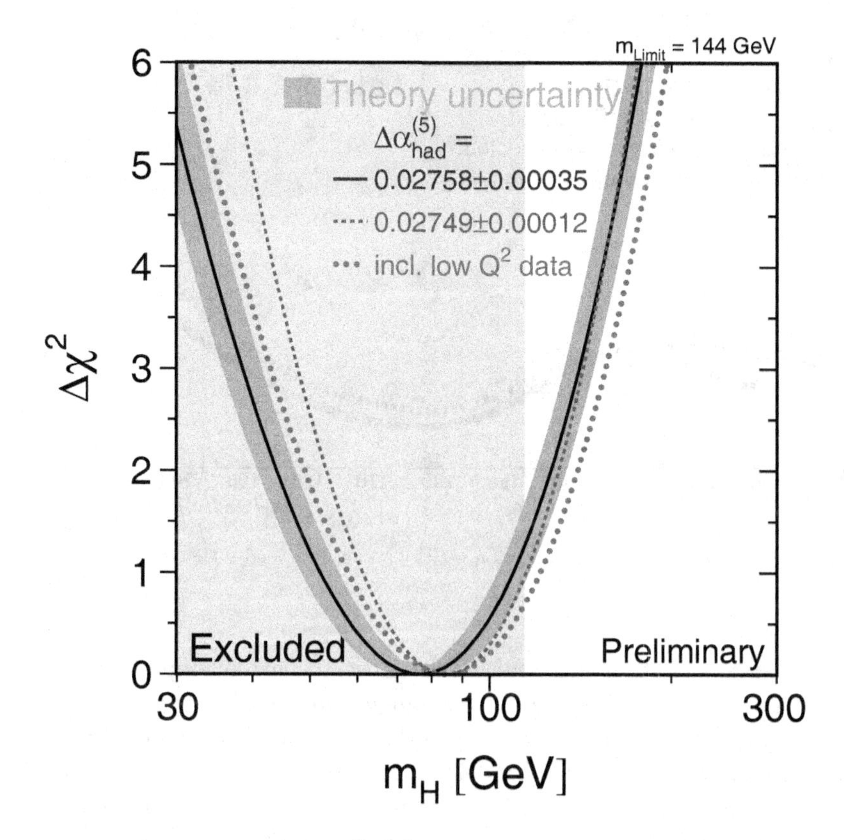

Fig. 2. Preferred SM Higgs mass.[5] $\Delta\chi^2$ curve derived from high-Q^2 precision electroweak measurements, performed at LEP and by SLD, CDF, and D0, as a function of the Higgs-boson mass, assuming the SM to be the correct theory of nature.

energy cut-off that sets in to maintain a finite self-coupling. A large cut-off results in a small Higgs mass (an energy cut-off of $10^{16} GeV$ results in a $m_{\mathrm{H}} \lesssim 200 GeV/c^2$.[17] A small cut-off energy allows a larger Higgs mass. A cut-off approximately equal to the Higgs mass results in an upper limit $m_{\mathrm{H}} \lesssim 700 GeV/c^2$. This *triviality* upper bound is thus similar to that from the perturbative unitarity requirement.

A theoretical lower bound on the Higgs mass is obtained from a requirement for stability of the vacuum ground state [see Ref. 6 and references therein]. The Higgs mass has a cut-off energy dependent lower bound which plateaus at a value of approximately $100 GeV/c^2$ for a cutoff energy above 1TeV. Thus relatively light Higgs masses are favoured, and certainly interesting new physics must show up below the TeV scale.

Considerable interest has been generated over recent years for theories to included supersymmetry, transforming fermions to bosons and vice-versa. As no evidence for supersymmetry exists to date, it is assumed to be a broken symmetry, with

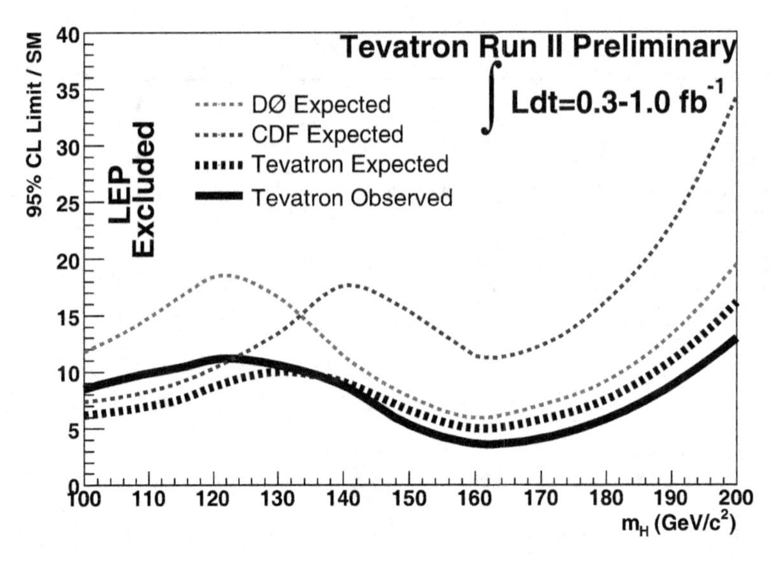

Fig. 3. Expected and observed 95% CL cross section ratios for the combined CDF and D analyses.[14]

arguments for the symmetry-breaking scale to be close to current experimental limits. Supersymmetric theories have the rather unique property of being absent of the quadratic quantum corrections that plague the SM. Such corrections result in the need for *unnatural* tuning of the model parameters when extrapolating to high energies. Supersymmetry thus naturally resolves the mass hierarchy problem.[6] Symmetry breaking schemes result in light Higgs masses just above current limits. Also the lightest supersymmetric particles (LSP), such as the *neutralino* is stable in many supersymmetry theories and is thus a candidate for the dark matter observed in astrophysical data.

There are other suggestions for solving the hierarchy problem, involving models with extra space dimensions, where some of these extra dimensions, at least, might become observable at the LHC. A possibility is that gravity is strong in the extra dimensions, perhaps resulting in the production of microscopic black holes at the TeV scale of the LHC. Microscopic black holes would decay very quickly via Hawking radiation, democratically emitting photons, neutrinos and other particles with spectacular signatures.[a]

4. The Large Hadron Collider

Within the framework of the SM and the minimal version of the SM Higgs field, the production rate [Fig. 4 and Ref. 8 and references therein] and decay modes and

[a]Amongst the broad range of interests the has been the hallmark of Bruce McKellar and GIrish Joshi is some work on black hole production.[7]

their branching ratios[17] [see Fig. 5] can all be calculated as a function of the Higgs mass.

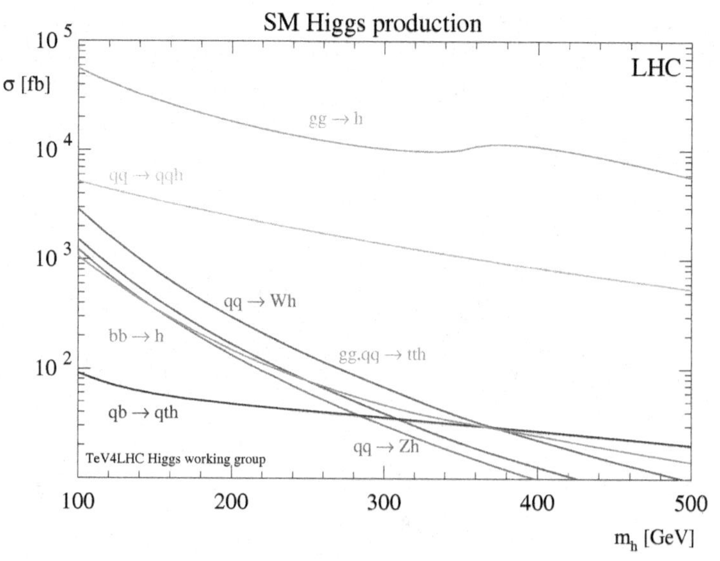

Fig. 4. SM Higgs production cross sections for the dominant processes in proton-proton interactions.[8]

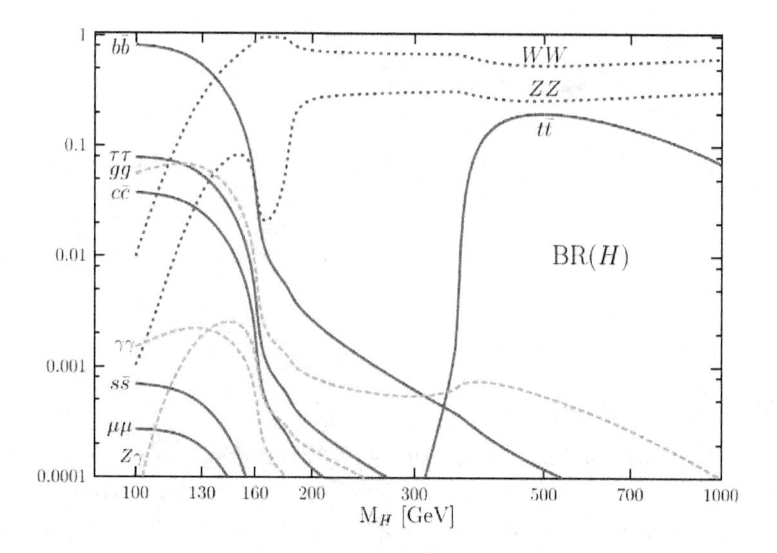

Fig. 5. SM Higgs Decay Branching Ratios.[17]

Whilst designing for a variety of possible new physics signatures, the LHC and the two general purpose experiments ATLAS and CMS were specifically benchmarked against sensitivity to the Higgs production and decay for a broad range of models. From the LHC machine perspective, the beam energy was limited by the available LEP tunnel radius, and by achievable magnet technology [see Fig. 6] . With a 14TeV beam CMS energy, the small cross section for Higgs production demanded a luminosity of $10^{34} cm^{-2} s^{-1}$ or higher. The inclusive cross section for all inelastic collisions is however, very high at these energies. Thus the LHC will have very high event rates, with high energy and multiplicity interactions, resulting in a difficult environment for the experiments.

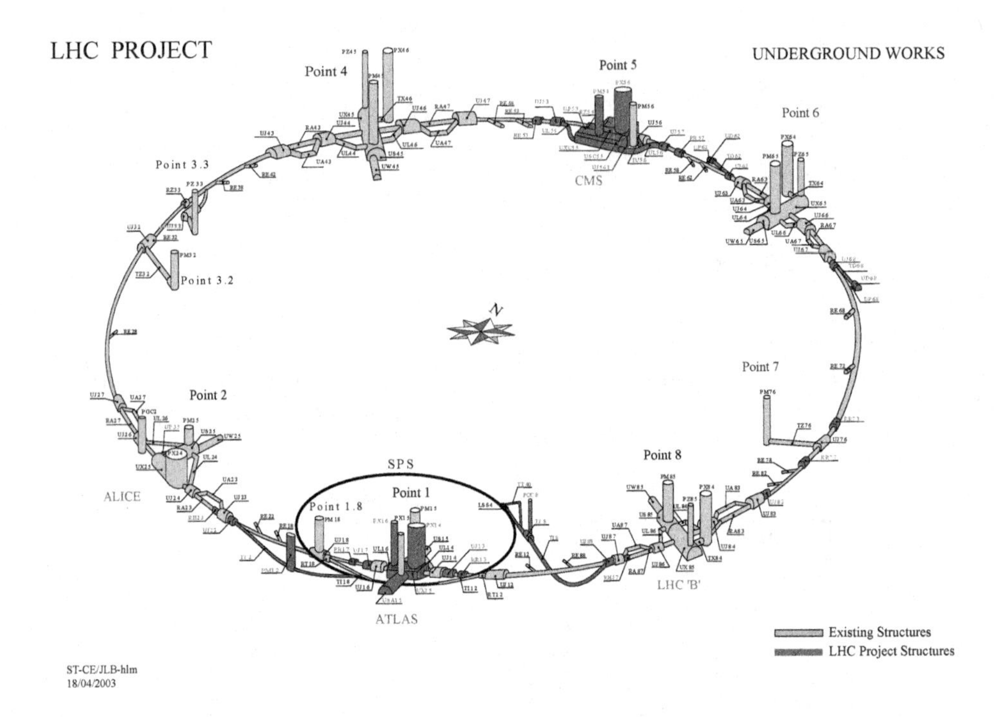

Fig. 6. Layout of the Large Hadron Collider.

5. The ATLAS Experiment

With Australian groups taking part in the ATLAS experiment[18] this paper will describe key features of this experiment and will direct the reader to the CMS literature to pursue differences between the two general purpose experiments, designed to exploit the discovery potential of the LHC.

The ATLAS experiment is shown schematically in Fig. 7
The detector optimization is primarily aimed at sensitivity to the largest possible Higgs mass range. Other important goals are the searches for heavy W- and Z-like

objects, and for supersymmetric particles. The design was aimed at capability to measured a broad variety of possible physics processes to maximize the detector's potential for the discovery of new, unexpected physics.

$H \to \gamma\gamma$	mass range $90 < m_H < 150$ GeV;
$H \to ZZ^* \to 4\ell^\pm$	mass range 130 GeV $< m_H < 2m_Z$;
$H \to ZZ \to 4\ell^\pm,\ 2\ell^\pm 2\nu$	mass range $m_H > 2m_Z$;
$H \to WW,\ ZZ \to \ell^\pm\nu$ 2 jets, $2\ell^\pm$ 2 jets	from WW, ZZ fusion using tagging of forward jets for m_H up to about 1 TeV.

Other examples of exploratory LHC physics which have been used to benchmark the detector design includes the signature of undetected lightest stable supersymmetric particles (LSP) for the study of supersymmetry (SUSY), where it is necessary to set stringent requirements for the hermeticity and E_T^{miss} capability of the detector. New, heavy, gauge bosons W′ and Z′ could be accessible to the LHC for masses up to several TeV, with excellent signatures in leptonic decay channels High-resolution lepton measurements are therefore needed out to $p_T$of TeV more more. FInally, signatures of new physics will be sought studying very high-p_T jets.

The basic design considerations for ATLAS can be summarized as:

- very good electromagnetic calorimetry for electron and photon identification and measurements, complemented by hermetic jet and missing $E_T$calorimetry;
- efficient tracking at high luminosity for lepton momentum measurements, for b-quark tagging, and for enhanced electron and photon identification, as well as tau and heavy-flavour vertexing and reconstruction capability of some B decay final states at lower luminosity;
- stand-alone, precision, muon-momentum measurements up to highest luminosity, and very low-p_T trigger capability at lower luminosity.

The ATLAS design, together with an expectation of the proton-proton luminosity from the LHC, is able to cover the complete range of Higgs masses from the current lower to the TeV range. Figire 8 shows the variety of decay channels that will be be capable of covering the entire range of masses for a 5-σ measurement of the SM Higgs.

6. First Measurements at the LHC

The current schedule (barring any unforeseen mishaps) for the LHC sees first physics beams in July 2008, with the aim of reaching a luminosity of $10^{32} cm^{-2} s^{-1}$ thus

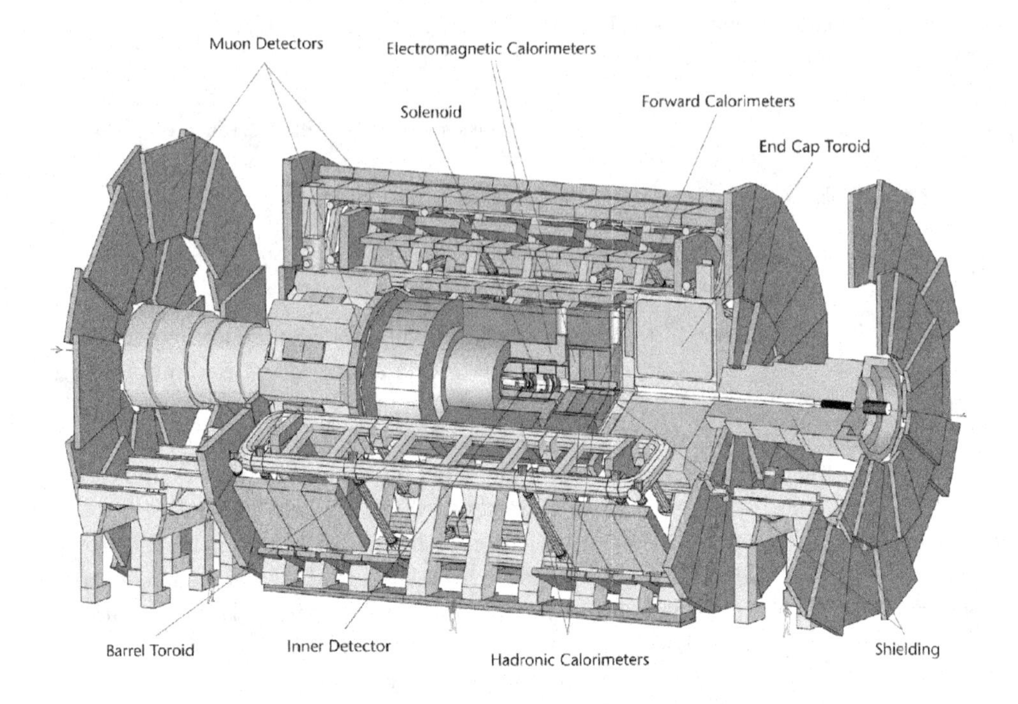

Fig. 7. The ATLAS Experiment.

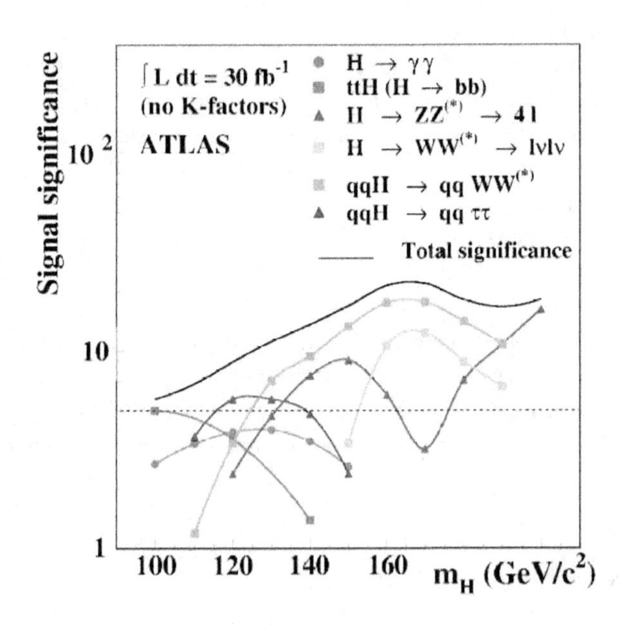

Fig. 8. The ATLAS discovery potential over the complete range of possible SM Higgs masses.

the collaborations will be hoping for some hundreds of pb^{-1} to $1fb^{-1}$ of integrated luminosity by the end of 2008. Generally, Higgs production cross sections decrease with Higgs mass, so with limited integrated luminosity, early searches will be limited to low Higgs masses ($m_{\rm H} \lesssim 200GeV/c^2$) except at very low masses ($m_{\rm H} \lesssim 130GeV/c^2$) where backgrounds to the most prominent signals become a major experimental challenge. Generally the Higgs has stronger coupling to the heaviest fermions (b-quarks and τ leptons), unless the mass of the Higgs is above threshold to decay into WW or ZZ pairs. However, hadronic final states of these particles will be swamped by the high cross section of hadron production at these energies. Final states with leptons or gammas will be required to make meaningful searches in the early statistics limited data. $H \rightarrow ZZ^* \rightarrow 4l$ and $H \rightarrow WW^* \rightarrow I\nu l\nu$ can be early discovery channels. However, even these channels will require $\sim 10fb^{-1}$ integrated luminosity, which will more than likely take 2009 data to achieve. It should be remembered, that understanding the detectors to a high degree of reliability and precision will be required to achieve the high efficiencies and discrimination implicit in the calculations of sensitivity for discovery. The early data will provide large samples of *known* physics for calibration and alignment. For example, $Z \rightarrow e^+e^-$ events will occur at rates of 5000 per day even at $10^{32}cm^{-2}s^{-1}$ luminosity. Such data will be essential for early calibration of the electromagnet calorimeter. Similarly, QCD jet production will be prolific, essential for hadronic calorimeter calibration and for understanding the background these processes represent to Higgs decay channels containing jets. Although low mass SM Higgs will take some time to discover, if supersymmetry exists even a $1TeV/c^2$ mass scale can be approached with $1fb^{-1}$ or more integrated luminosity, via cascade processes, with missing energy. Another signal that will be easily observed in early data taking is a high mass Z' that decays into two leptons. Such particles with masses in the $2TeV/c^2$ range will be significant with only a few hundred pb^{-1} integrated luminosity.

It should also be noted that the LHC will produce copious B-mesons including B_s-mesons. ATLAS and CMS will have active B-physics programs, alongside the dedicated experiment, LHCb.[20] Over the last few years, experiments at the KEK (*Belle*) and SLAC (*BaBar*) B-meson factories have measured mixing of neutral B-mesons with anti-B-mesons with high precision, have studied CP-violation and rare decays. Consistency of the experiments with CKM quark-mixing in the SM has withstood the increased precision of the experiments. Figure 9 shows the current fit of data from these experiments and others in the CKM matrix unitarity plane. The *Unitarity Triangle* being consistently mapped out with precision data is yet another testament to the SM. Searches for new physics at the TeV scale could reveal an explanation of the matter-antimatter in the universe, that has eluded Belle and Babar. [b]

[b] Bruce McKellar and Girish Joshi have published in the area B-physics for many years [eg. Ref. 9] connecting with their Australian colleagues on the Belle experiment.

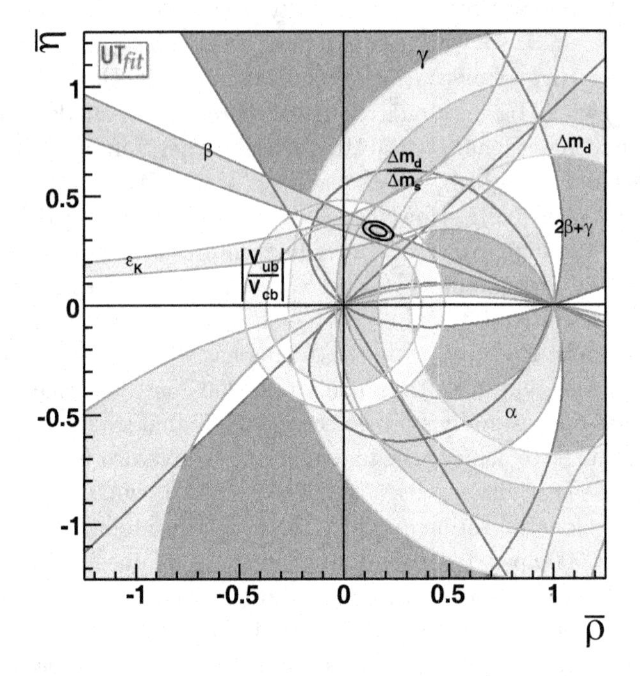

Fig. 9. $\rho - \eta$ plot: CKM Unitarity Triangle.[13]

7. The Outlook

Particle physics is entering a new era with excellent prospects exist for discovering new physics beyond the SM. The LHC has the potential to revolutionize our understanding of particle physics and the very space we live in. To reach the design luminosity with high-quality beam delivered to the experiments will take some years. A one-off facility of this complexity and size was never going to be easy to fully exploit. However, physicists are already preparing for upgrades in luminosity that are expected to fully reap the fruits of the investment that the LHC and experiments represent. Another factor of 10 over the design luminosity is talked about. Such an increase will required upgrades to the detector capability as well as the accelerator. Such improvements will require years of experience with the current equipment, but with decade-long lead-times required, the early upgrade investigations are not premature. Furthermore, with the LHC described as a discovery machine, the community is seriously working on a complementary electron-positron collider with CMS energy in the $500GeV - 1TeV$ range, planned to take up the challenge of precision measurements of the Tera-scale physics that the LHC will uncover. The International Linear Collider (ILC) will not have its final specifications until operation at LHC has fully define the target energy range and broad spectrum of new particles or phenomena. The need for a high precision machine is based upon past experience where the power hadron machines, fixed target and then colliders, in pushing through new frontiers, had to be matched by the finesse of lepton scattering and electron-positron colliders, to map out the subtile secrets of nature. The scope of the ILC is known even before final specifications to be of such proportions and so far beyond present experience, that the technology and engineering development projects for the ILC are already underway.

References

1. D. N. Spergel, et.al. (WMAP Collaboration), *ApJS* , **170**, 377 (2007)
2. S. Glashow, *Nucl. Phys.* **22**, 579 (1961)
3. S. Weinberg, *Phys. Rev. Lett.* **19**, 1264 (1967)
4. A. Salam, in *Elementary Particle Physics; Relativistic Groups and Analyticity* (Nobel Symposium No. 8) (ed. N. Svartholm) 367 (Almqvist and Wiksills, Stockholm, 1968).
5. LEP Electroweak Working Group http://lepewwg.web.cern.ch/LEPEWWG (2007)
6. J. Ellis, G. Ridolfi, F. Zwirner, *C. R. Physique*, in press (2007), arXiv:hep-ph/0702114v1
7. S. R. Choudhury, A. S. Cornell, G. C. Joshi, B. H. J. McKellar, *Mod. Phys. Lett.* **A19**, 2331 (2004)
8. http://maltoni.home.cern.ch/maltoni/TeV4LHC/ (2007)
9. S. Rai Choudhury, A. S. Cornell, G. C. Joshi, B. H. J. McKellar, *Phys. Rev.*,**D74**, 054031(2006)
10. P. W. Higgs *Phys. Rev. Lett.* **13**, 508 (1964)
11. A. A. Aguilar-Arevalo, et al. (The MiniBooNE Collaboration) *Phys. Rev. Lett.* **98** 231801 (2007).
12. F. J. Hasert, et al. (Gargamelle Collaboration) *Phys. Lett.* **46B**, 121 (1973)
13. UTfit collaboration *arXiv:hep-ph/0606167v2*, (2006) and http://utfit.roma1.infn.it
14. Tevatron New-Phenomena and Higgs Working Group, http://tevnphwg.fnal.gov (2007)
15. Tevatron Electroweak Working Group *arXiv:hep-ex/0703034* (2007)
16. M. J. G. Veltman, *Acta Phys. Polon.*, **B8**, 475 (1977)
17. A. Djouadi *arXiv:hep-ph/0503172v2*, (2005)
18. The ATLAS Collaboration, http://atlas.web.cern.ch/Atlas/GROUPS/PHYSICS/TDR/access.html
19. The CMS Collaboration, CERN-LHCC-2006-021 (2006)
20. The LHCb Collaboration, http://lhcb.wed.cern.ch/lhcb/ (2007)

SOME RECENT LATTICE QCD RESULTS FROM THE CSSM

S. BOINPOLLI[1], P. O. BOWMAN[2], J. N. HEDDITCH[1], U. M. HELLER[3], W. KAMLEH[1], B. G. LASSCOCK[1], D. B. LEINWEBER[1], A. G. WILLIAMS[1,†], J. M. ZANOTTI[4] and J. B. ZHANG[5]

[1] *Special Research Centre for the Subatomic Structure of Matter and Department of Physics, The University of Adelaide, 5005, Australia*
[2] *Centre of Theoretical Chemistry and Physics, Institute of Fundamental Sciences, Massey University (Auckland), Private Bag 102904, NSMSC, Auckland NZ*
[3] *American Physical Society, One Research Road, Box 9000, Ridge, NY 11961-9000, USA*
[4] *School of Physics, University of Edinburgh, Edinburgh EH9 3JZ, UK*
[5] *Department of Physics, Zhejiang University, Hangzhou, Zhejiang 310027, P.R. China*
* *Email: anthony.williams@adelaide.edu.au*

Recent CSSM Lattice Collaboration studies of hadron electromagnetic structure and the exotic hadron spectrum are highlighted. The momentum dependence of quark and gluon propagators revealed in lattice simulations of full QCD are also illustrated.

Keywords: Lattice QCD; Electromagnetic Form Factors, Quark and Gluon Propagators.

1. Introduction

The CSSM Lattice Collaboration is active in a variety of research areas sharing a common goal of furthering the understanding of strong interaction physics. Core to this research program are the studies of hadron structure and the hadron spectrum using the FLIC fermion action.[1] This action provides an alternative form of nonperturbative $\mathcal{O}(a)$ improvement[2] providing excellent scaling.[3] Its improved chiral properties[4] provide efficient access to the chiral regime enabling large volume studies.

The important role that electromagnetic form factors play in our understanding of hadronic structure has been well documented for more than fifty years. The reason for their popularity is that they encode information about the shape of hadrons, and provide valuable insights into their internal structure in terms of quark and gluon degrees of freedom. Moreover baryon charge radii and magnetic moments provide an excellent opportunity to observe the chiral non-analytic quark-mass dependence of QCD observables. In Sec. 2 we present some of the results of our latest precision study into the electromagnetic structure of octet baryons.[5]

[†] Speaker.

Early CSSM Lattice Collaboration studies of the hadron spectrum have focused on the excited states of conventional baryons,[6] with the most recent study focusing on the spectrum of even parity excited states of the nucleon[7] in a search for the Roper resonance. However in recent years, experiment has revealed putative evidence of the existence of the more exotic hadrons. On the lattice we have a unique opportunity to probe for the existence of these exotic states in QCD from first principles simulations. Specifically we searched for the existence of the Θ^+ pentaquark[8,9] and the $J^{PC} = 1^{-+}$ exotic meson.[10] The states are considered exotic because, although not apparently forbidden in QCD, these states are not described in terms of the simple quark model. In Sec. 3, we present the results of our study into the 1^{-+} exotic meson.

Finally, it is possible in lattice QCD to probe aspects of QCD not directly accessible in experiment. Of particular interest is the nature of flux-tube formation in two- and three-quark systems.[11,12] Similarly, it is important to learn the impact of dynamical sea-quarks on the nature of quark and gluon propagators.[13–18] Key aspects of our analysis of quark and gluon propagators in quenched and full lattice QCD are presented in Sec. 4. Finally in Sec. 5, we discuss the directions of future CSSM lattice studies.

2. Electromagnetic Structure of Octet and Decuplet Baryons

Research into the electromagnetic structure of baryons has been a primary focus of the CSSM,[5,19–21] with more recent studies extending this work to the study of pseudoscalar and vector meson form factors.[22] In this section, we present the results of our precision study of the electromagnetic structure of octet baryons.[5]

In lattice QCD, information regarding the hadron mass spectrum and structure is encoded within two- and three-point correlation functions. For instance we can extract the ground state hadron masses in lattice QCD by fitting the slope of the log of the two-point correlation function (projected to zero momentum) in the asymptotic limit of large Euclidean time. For baryons the two-point correlation function in momentum space is

$$\langle\, G^{BB}(t;\vec{p},\Gamma_\pm)\,\rangle = \sum_{\vec{x}} e^{-i\vec{p}\cdot\vec{x}}\,\Gamma_\pm\,\langle\,\Omega\,|\,T\,(\chi(x)\,\overline{\chi}(0))\,|\,\Omega\,\rangle\,, \tag{1}$$

where the interpolator $\chi(\bar{\chi})$ annihilates (creates) baryon states to (from) the vacuum, and $\Gamma_\pm$ is the parity projection operator[23] (the $+$ and $-$ label even and odd parity respectively). A review of the octet baryon interpolating fields we considered can be found in Ref. 24. For a complete derivation of how baryon masses are extracted from the two-point correlation function see Refs. 7, 24.

The matrix element describing the interaction of a hadron with a vector current contains information regarding the hadron's structure. This matrix element for octet

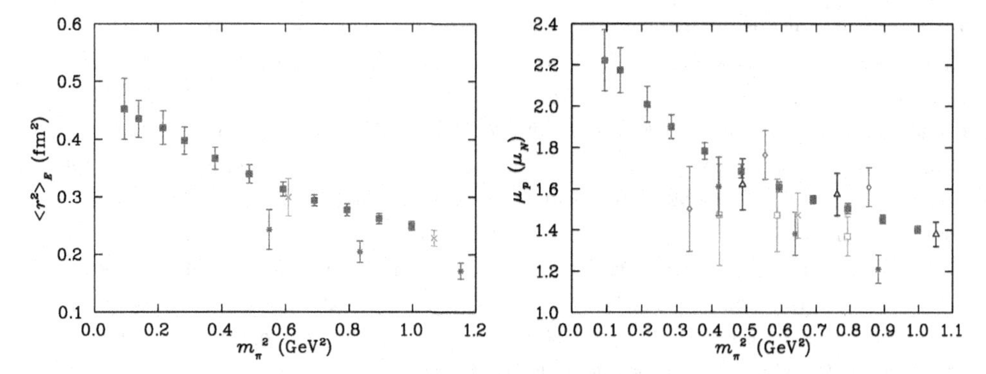

Fig. 1. (left) Proton charge radius (see text).[26] (right) Proton magnetic moment (in nuclear magnetons) compared with a variety of lattice simulations (see text).[26]

baryons is

$$\langle\, p', s' \mid j^{\mu} \mid p, s \,\rangle = \left(\frac{M^2}{E_p E_{p'}}\right)^{1/2} \overline{u}(p', s') \left(F_1(q^2)\gamma^{\mu} - F_2(q^2)\sigma^{\mu\nu}\frac{q^{\nu}}{2M}\right) u(p, s)\,, \tag{2}$$

where $q = p' - p$. This matrix element appears in the three-point correlation function

$$\begin{aligned} &G^{Bj^{\mu}B}(t_2, t_1; \vec{p'}, \vec{p}; \Gamma) \\ &= \sum_{\vec{x_2}, \vec{x_1}} e^{-i\vec{p'}\cdot\vec{x_2}} e^{+i(\vec{p'}-\vec{p})\cdot\vec{x_1}} \Gamma \,\langle\, \Omega \mid T\left(\chi(x_2) j^{\mu}(x_1) \overline{\chi}(0)\right) \mid \Omega \,\rangle\,, \end{aligned} \tag{3}$$

when it is evaluated at the phenomenological level by introducing complete sets of baryon intermediate states. We extract the form factors $F_1(q^2)$ and $F_1(q^2)$ by taking ratios of the two- and three-point functions (Eqs. (1) and (3)). The charge radii and magnetic moments of the octet baryons are extracted from the Sachs form factors, which are related to F_1 and F_2 by,

$$\mathcal{G}_E(q^2) = F_1(q^2) - \frac{q^2}{(2M)^2} F_2(q^2)\,, \qquad \mathcal{G}_M(q^2) = F_1(q^2) + F_2(q^2)\,. \tag{4}$$

For more details see Ref. 25.

Lattice simulations are performed using the mean-field $O(a^2)$-improved Luscher–Weisz plaquette plus rectangle gauge action on a $20^3 \times 40$ lattice with periodic boundary conditions. The lattice spacing is $a = 0.128$ fm as determined by the Sommer scale $r_0 = 0.50$ fm. The reasonably large physical volume lattice ensures a good density of low-lying momenta which are key to giving rise to chiral non-analytic behavior in the observables simulated on the lattice. A total of 400 configurations are considered, with error analysis by third-order, single elimination jackknife. As discussed in Sec. 1 we use the FLIC fermion action[1–4] with a smeared source and point sink. Our definition of the vector current is both conserved and $\mathcal{O}(a)$-improved. Complete simulation details are described in Ref. 5.

Our simulations of the proton charge radius are shown in Fig. 1 (left) where the result is compared with the previous state of the art lattice simulation results. The solid squares in this figure indicate current lattice QCD results with FLIC fermions. The stars indicate the lattice results of Leinweber *et. al.*[25] while the crosses indicate the results of Wilcox *et. al.*[27] both of which use the standard Wilson actions for the gauge and fermion fields. The results of simulations of the proton magnetic moment are shown in Fig. 1 (right) where the solid squares indicate our current lattice QCD results with FLIC fermions, the stars indicate the results of Leinweber *et. al.*,[25] crosses (only one point) for Wilcox *et. al.*,[27] open symbols for the QCDSF collaboration results, Gockeler *et. al.*[28] (where open squares correspond to $\beta = 6.0$, open triangles to $\beta = 6.2$, open diamonds to $\beta = 6.4$).

The improved chiral properties of the FLIC fermion action provides good access to the chiral regime and, for the first time, reveals chiral curvature in accord with the expectations of quenched chiral effective field theory.[29]

3. Hybrid and Exotic Mesons from FLIC Fermions

As discussed in Sec. 1, exotic mesons are mesons with quantum numbers which cannot be carried by a simple q-$\bar{q}$ pair. Theoretical studies of these exotic states are timely as major experimental efforts are planned in the 12 GeV Hall D upgrade at Jefferson Lab.

To create these mesonic states with exotic quantum numbers in lattice QCD, non-local, hybrid or four-quark interpolators are possible avenues. In our study of the 1^{-+} channel,[30] we gain access to these states with local hybrid-meson interpolators which include explicit chromo-electric and -magnetic degrees of freedom. In Ref. 30, our complete basis of local meson interpolating fields for the each set of quantum numbers is summarized. Of the four interpolators that access states with the 1^{-+} quantum numbers, we consider the two interpolators

$$\begin{aligned}\chi_2 &= i\epsilon_{jkl}\bar{q}^a \gamma_k B_l^{ab} q^b \\ \chi_3 &= i\epsilon_{jkl}\bar{q}^a \gamma_4 \gamma_k B_l^{ab} q^b\,,\end{aligned} \tag{5}$$

both of which couple the *large* $\times$ *large* components of the spinors. The analysis of our results extracted with the χ_2 interpolator are presented here.

In Fig. 2 (top) we show a summary of the 1^{-+} meson masses calculated both from quenched and full QCD simulations plotted as a function of the squared pion mass. The MILC collaboration data is taken from Bernard et. al.,[31,32] and the SESAM data is taken from Lacock et. al.[33] Note that open and closed symbols denote dynamical and quenched simulations respectively. Our results compare favorably with earlier work at large quark masses. Agreement within one sigma is observed for all the quenched simulation results illustrated by filled symbols.

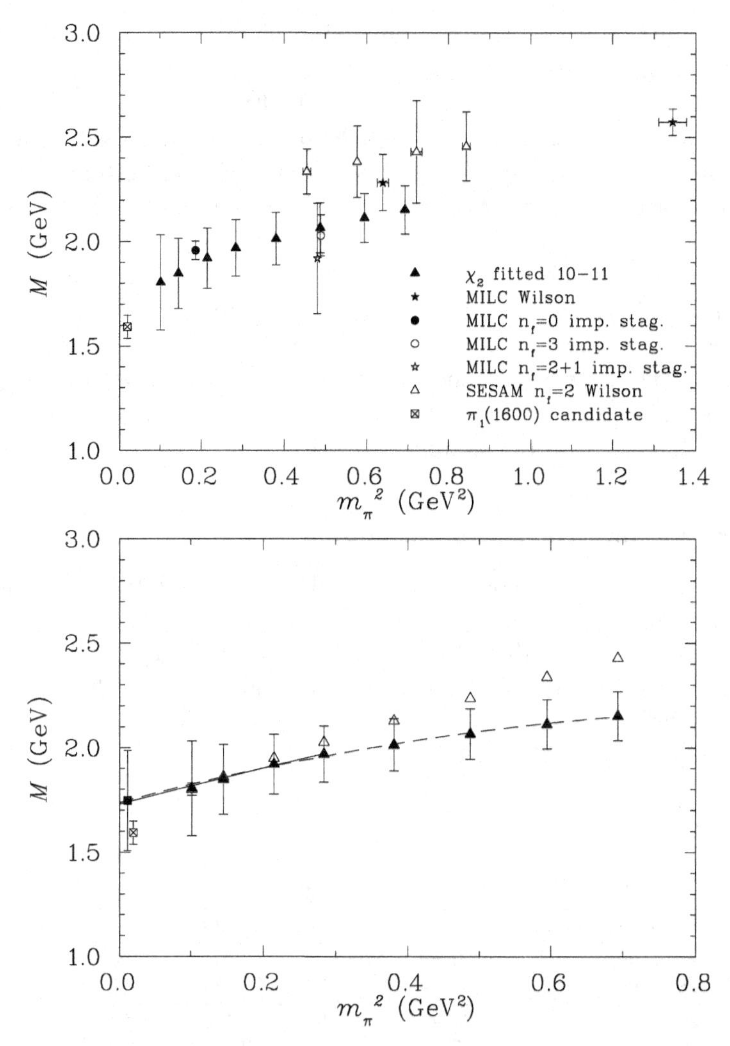

Fig. 2. (top) A survey of lattice results for the 1^{-+} meson mass.[30] (bottom) The 1^{-+} exotic meson mass obtained from the hybrid interpolator χ_2 (full triangles) are compared with the $a_1\eta'$ two-particle state energy (open triangles).[30]

The particle data group reports two experimental candidates for the 1^{-+} exotic meson, the $\pi_1(1400)$ and the $\pi_1(1600)$. To compare our lattice results to experiment, it is necessary to extrapolate our results to the the physical pion mass. In Fig. 2 (bottom) we show the line of best fit of a function that is quadratic in m_π^2 at all eight quark masses (dashed line) and a fit linear in m_π^2 through the four lightest quark masses (solid line). The full square is result of linear extrapolation to the physical pion mass, while the open square (offset for clarity) indicates the $\pi_1(1600)$ experimental candidate. For a discussion of the systematic errors in this extrapolation see Ref. 30.

Fig. 2 (bottom) illustrates how our extrapolated results favour the existence of the $\pi_1(1600)$ exotic meson state. However our interpolators couple to all possible states with the appropriate quantum numbers, including multi-hadron states. In quenched lattice QCD the η' is degenerate in mass with the pion. This gives rise to a low lying $a_1\eta'$ multi-hadron state in the spectrum, also shown in Fig. 2 (bottom). Therefore we have to consider to what extent the contribution from this multi-hadron state is contaminating our results. Chiral perturbation theory suggests that the $a_1\eta'$ channel should come with a negative metric contribution to the correlation function. Since our correlation functions are positive, it suggests that coupling to this state is small. Further, at the larger quark masses the correlation functions are dominated by a bound 1^{-+} state (i.e. lighter than the lowest energy multi-hadron state) at small euclidean times. Therefore we conclude that our interpolators are well suited to extracting the single-particle 1^{-+} state.

4. Quark and Gluon Propagators in Full QCD

In this section, we present an unquenched calculation of the quark propagator in Landau gauge with $2+1$ flavors of dynamical quarks. We study the scaling behavior of the quark propagator in full QCD on two lattices with different lattice spacings and similar physical volume. We use configurations generated with an improved staggered ("AsqTad") action by the MILC collaboration.[34] The lattice parameters used in this study are shown in Table 1.

Table 1. Lattice parameters used in this study. The dynamical configurations each have two degenerate light quarks (up/down) and a heavier quark approximating the strange quark mass.

Dimensions	β	a	Bare Quark Mass	Config
$28^3 \times 96$	7.09	0.090 fm	14.0 MeV, 67.8 MeV	108
$28^3 \times 96$	7.11	0.090 fm	27.1 MeV, 67.8 MeV	110
$20^3 \times 64$	6.76	0.125 fm	15.7 MeV, 78.9 MeV	203
$20^3 \times 64$	6.79	0.125 fm	31.5 MeV, 78.9 MeV	249
$20^3 \times 64$	6.81	0.125 fm	47.3 MeV, 78.9 MeV	268
$20^3 \times 64$	6.83	0.125 fm	63.1 MeV, 78.9 MeV	318

A comparison of quenched and full QCD is displayed in Fig. 3. Full triangles correspond to the quenched calculation, while open circles correspond to the $2+1$ flavor QCD. As the lattice spacing and volume are the same, the difference between the two results is entirely due to the presence of sea-quark loops. The mass function for the unquenched dynamical-fermion propagator has been interpolated so that it agrees with the quenched mass function for $ma = 0.01$ at the renormalization point, $q = 3$ GeV. For the unquenched propagator this corresponds to a bare quark mass of $ma = 0.0087$. Note that the effect of unquenching is to move propagators toward their tree-level form, *i.e.* more Abelian-like as expected naively.

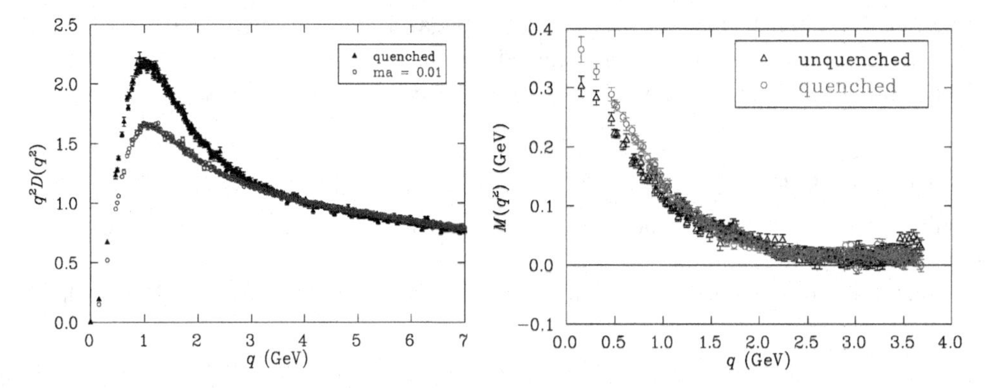

Fig. 3. The gluon dressing function in Landau gauge (left) and the quark propagator mass function for non-zero quark mass (right) in quenched and dynamical-fermion QCD.[35]

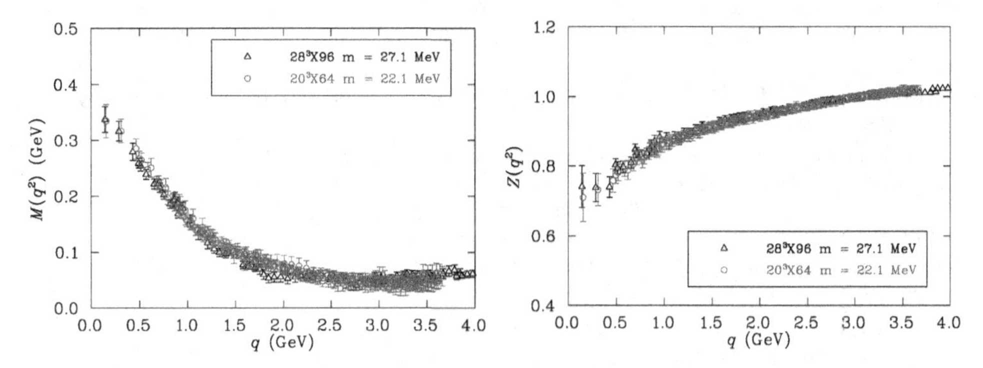

Fig. 4. (left) Scaling of the quark propagator mass function $M(q^2)$ in full QCD.[36] (right) Scaling of the wave-function renormalisation function $Z(q^2)$ in full QCD.[36]

We also investigate the scaling of the quark propagator for full QCD, as shown in Fig. 4, where the triangles correspond to a $28^3 \times 96$ lattice with spacing $a = 0.009$ fm, while the open circles are for a $20^3 \times 64$ lattice with spacing $a = 0.125$ fm (obtained by interpolating four different sets of light quark masses). These results are for two different $2+1$ flavor MILC lattices renormalized to agree at 3 GeV and with similar volumes. The results illustrated in Fig. 4 display good scaling behavior.

5. Conclusions

This is an interesting and exciting time to be engaged in the numerical simulation of QCD. Years of lattice action improvement, algorithmic development and a steady increase in supercomputing capability mean that steady growth in both the precision of QCD predictions and the understanding of QCD mechanisms is underway now. The CSSM lattice collaboration is contributing to this success in several areas including explorations of the electromagnetic structure of baryons, quenched artifacts, chiral curvature, quark environment sensitivity, hybrid and exotic meson studies, and unquenched quark and gluon propagator calculations. Future studies will continue to focus more on unquenched dynamical-FLIC simulations[37] which hold the promise of providing efficient access to the chiral regime of Nature.

Acknowledgments

We give our congratulations and best wishes to Bruce McKellar and Girish Joshi. We are grateful for grants of supercomputer time from SAPAC and APAC which have enabled this project. This research is supported by the Australian Research Council.

References

1. J. M. Zanotti *et al.*, *Phys. Rev.* **D65**, p. 074507 (2002).
2. D. B. Leinweber *et al.*, *nucl-th/0211014* (2002).
3. J. M. Zanotti, B. Lasscock, D. B. Leinweber and A. G. Williams, *Phys. Rev.* **D71**, p. 034510 (2005).
4. S. Boinepalli, W. Kamleh, D. B. Leinweber, A. G. Williams and J. M. Zanotti, *Phys. Lett.* **B616**, 196 (2005).
5. S. Boinepalli, D. B. Leinweber, A. G. Williams, J. M. Zanotti and J. B. Zhang, *Phys. Rev.* **D74**, p. 093005 (2006).
6. W. Melnitchouk *et al.*, *Nucl. Phys. Proc. Suppl.* **119**, 293 (2003).
7. B. G. Lasscock *et al.*, *hep-lat/0705.0861* (2007).
8. B. G. Lasscock *et al.*, *Phys. Rev.* **D72**, p. 014502 (2005).
9. B. G. Lasscock *et al.*, *Phys. Rev.* **D72**, p. 074507 (2005).
10. J. N. Hedditch *et al.*, *Phys. Rev.* **D72**, p. 114507 (2005).
11. F. Bissey *et al.* (2006).
12. F. Bissey *et al.*, *Nucl. Phys. Proc. Suppl.* **141**, 22 (2005).
13. F. D. R. Bonnet, P. O. Bowman, D. B. Leinweber, A. G. Williams and J. M. Zanotti, *Phys. Rev.* **D64**, p. 034501 (2001).
14. P. O. Bowman, U. M. Heller, D. B. Leinweber, M. B. Parappilly and A. G. Williams, *Phys. Rev.* **D70**, p. 034509 (2004).
15. P. O. Bowman *et al.*, *Phys. Rev.* **D71**, p. 054507 (2005).
16. W. Kamleh, P. O. Bowman, D. B. Leinweber, A. G. Williams and J. B. Zhang, *Nucl. Phys. Proc. Suppl.* **161**, 109 (2006).
17. W. Kamleh, P. O. Bowman, D. B. Leinweber, A. G. Williams and J. Zhang (2007).
18. P. O. Bowman *et al.*, *hep-lat/0703022* (2007).
19. J. M. Zanotti, S. Boinepalli, D. B. Leinweber, A. G. Williams and J. B. Zhang, *Nucl. Phys. Proc. Suppl.* **128**, 233 (2004).
20. D. B. Leinweber *et al.*, *Phys. Rev. Lett.* **94**, p. 212001 (2005).
21. D. B. Leinweber *et al.*, *Phys. Rev. Lett.* **97**, p. 022001 (2006).
22. J. N. Hedditch *et al.*, *Phys. Rev.* **D75**, p. 094504 (2007).
23. F. X. Lee and D. B. Leinweber, *Nucl. Phys. Proc. Suppl.* **73**, 258 (1999).
24. W. Melnitchouk *et al.*, *Phys. Rev.* **D67**, p. 114506 (2003).
25. D. B. Leinweber, R. M. Woloshyn and T. Draper, *Phys. Rev.* **D43**, 1659 (1991).
26. S. Boinepalli, D. B. Leinweber, A. G. Williams, J. M. Zanotti and J. B. Zhang, *Phys. Rev.* **D74**, p. 093005 (2006).
27. W. Wilcox, T. Draper and K.-F. Liu, *Phys. Rev.* **D46**, 1109 (1992).
28. M. Gockeler *et al.*, *Phys. Rev.* **D71**, p. 034508 (2005).

29. D. B. Leinweber, *Phys. Rev.* **D69**, p. 014005 (2004).
30. J. N. Hedditch *et al.*, *Phys. Rev.* **D72**, p. 114507 (2005).
31. C. W. Bernard *et al.*, *Phys. Rev.* **D56**, 7039 (1997).
32. C. Bernard *et al.*, *Nucl. Phys. Proc. Suppl.* **119**, 260 (2003).
33. P. Lacock and K. Schilling, *Nucl. Phys. Proc. Suppl.* **73**, 261 (1999).
34. C. W. Bernard *et al.*, *Phys. Rev.* **D64**, p. 054506 (2001).
35. M. B. Parappilly *et al.*, *AIP Conf. Proc.* **842**, 237 (2006).
36. M. B. Parappilly *et al.*, *Phys. Rev.* **D73**, p. 054504 (2006).
37. W. Kamleh, D. B. Leinweber and A. G. Williams, *Phys. Rev.* **D70**, p. 014502 (2004).
38. S. Boinepalli *et al.*, *hep-lat/0611028* (2006).